Understanding Bolometers

Understanding Bolometers

Edited by **Bob Tucker**

New York

Published by NY Research Press,
23 West, 55th Street, Suite 816,
New York, NY 10019, USA
www.nyresearchpress.com

Understanding Bolometers
Edited by Bob Tucker

International Standard Book Number: 978-1-63238-459-1 (Hardback)

Printed in the United States of America.

Contents

Preface   VII

Part 1   Bolometer Materials   1

Chapter 1   **Un-Cooled Microbolometers with
Amorphous Germanium-Silicon (a-Ge$_x$Si$_y$:H)
Thermo-Sensing Films**   3
Mario Moreno, Alfonso Torres, Roberto Ambrosio
and Andrey Kosarev

Chapter 2   **Group IV Materials for Low Cost
and High Performance Bolometers**   31
Henry H. Radamson and M. Kolahdouz

Part 2   Bolometer Types and Properties   51

Chapter 3   **Cold-Electron Bolometer**   53
Leonid S. Kuzmin

Chapter 4   **Lens-Antenna Coupled
Superconducting Hot-Electron Bolometers for
Terahertz Heterodyne Detection and Imaging**   83
Lei Liu

Chapter 5   **Noise Limitations of
Miniature Thermistors and Bolometers**   111
Béla Szentpáli

Part 3   Advances and Trends   135

Chapter 6   **Bolometers for Fusion Plasma Diagnostics**   137
Kumudni Tahiliani and Ratneshwar Jha

Chapter 7   **Smart Bolometer: Toward**
             Monolithic Bolometer with Smart Functions   **157**
             Matthieu Denoual, Olivier de Sagazan,
             Patrick Attia and Gilles Allègre

Chapter 8   **Detection of Terahertz Radiation**
             from Submicron Plasma Waves Transistors   **183**
             Y. M. Meziani, E. Garcia, J. Calvo, E. Diez,
             E. Velazquez, K. Fobelets and W. Knap

             Permissions

             List of Contributors

Preface

A bolometer is a sensitive electrical instrument for measuring radiant energy. Infrared detectors and related technologies have important applications not only in defense sector, but also for distinct civilian purposes. Some cost effective modern devices based on infrared technologies have been rapidly emerging on the global front. Different facets of bolometer advancements have been described in this book. It encompasses materials used, types of bolometers, their performance, utilization and prospective trends. This book will be of great help not just to the experts and professionals but also to the students who are involved in this discipline of science.

This book is a comprehensive compilation of works of different researchers from varied parts of the world. It includes valuable experiences of the researchers with the sole objective of providing the readers (learners) with a proper knowledge of the concerned field. This book will be beneficial in evoking inspiration and enhancing the knowledge of the interested readers.

In the end, I would like to extend my heartiest thanks to the authors who worked with great determination on their chapters. I also appreciate the publisher's support in the course of the book. I would also like to deeply acknowledge my family who stood by me as a source of inspiration during the project.

Editor

Part 1

Bolometer Materials

Un-Cooled Microbolometers with Amorphous Germanium-Silicon (a-Ge$_x$Si$_y$:H) Thermo-Sensing Films

Mario Moreno[1], Alfonso Torres[1], Roberto Ambrosio[2] and Andrey Kosarev[1]
[1]National Institute of Astrophysics, Optics and Electronics, INAOE,
[2]Universidad Autonoma de Ciudad Juarez, UACJ,
Mexico

1. Introduction

Silicon integrated circuits (IC) in conjunction with the micro-machining technology for thin films, have opened new ways for the development of low cost and reliable night vision systems based on thermal detectors. Among the thermal detectors used as pixels on IR focal plane arrays, the microbolometer appears as one of them. A microbolometer is a device in which the IR transduction is performed through a change in the resistivity of its thermo-sensing material, due to the heating effect caused by the absorbed radiation. Among the requirements for the materials used as thermo-sensing layer in microbolometers it can be mentioned a high activation energy (E_a), high temperature coefficient of resistance (TCR), low noise, and compatibility with standard CMOS fabrication processes. A variety of materials have been used as thermo-sensing elements in microbolometers, as vanadium oxide (VO$_x$) (B. E. Cole, 1998, 2000), metals (A. Tanaka, 1996), polycrystalline (S. Sedky, 1998) and amorphous semiconductors (A. J. Syllaios, 2000).

Those materials have shown good characteristics but also some disadvantages. VO$_x$ has a moderated value of TCR (0.021 K^{-1}) and low resistivty, however it is not a standard material in the IC technology. Metals as titanium are compatible with the standard IC technology, have low resistivity but also have very low TCR values. Polycrystalline semiconductors have high TCR values (0.05 K^{-1}) and moderated resistivity, however they are deposited a relatively high temperatures (700 - 900 °C), which results in an incompatibility with a micro-bolometer fabrication post-process on a silicon wafer surface, containing an readout integrated circuit (ROIC).

Recently, it has been reported the study of W-doped VO$_2$ (H. Takami, 2011) which has a TCR of above 0.1 k^{-1}, and low resistivity values. However, this material is not standard on Si CMOS microelectronics facilities. GaAs/AlGaAs heterojunction bolometers (P.K.D.D.P. Pitigala, 2011) also have been reported, which have demonstrated TCR values of 0.04 K^{-1}. However those structures are very complex, since they are fabricated with 30 periods of GaAs/Al$_{0.57}$Ga$_{0.43}$As junctions.

Hydrogenated amorphous silicon (a-Si:H) is a mature material on the microelectronics and photovoltaic industries. For un-cooled microbolometers a-Si:H is very attractive to be used as thermo-sensing material, since intrinsic a-Si:H has a very large activation energy (E_a) of above 1 eV, and therefore, provide a very large thermal coefficient of resistance (TCR) of 0.13 K^{-1}. However intrinsic a-Si:H has a very low room temperature conductivity ($\sigma_{RT} \leq 1x10^{-9}$ $(\Omega cm)^{-1}$), resulting in a very high pixel resistance when is used as thermo sensing material in microbolometers ($R_{pixel} \geq 10^9\ \Omega$). Such high pixel resistance causes a mismatch with the input impedance of the CMOS ROIC. For commercial microbolometers, boron doping is commonly used in order to decrease the undesirable resistivity of intrinsic a-Si:H (A. J. Syllaios, 2000), to values of pixel resistance of around $30x10^6\ \Omega$, however it also results on a reduction on the activation energy ($E_a \approx 0.22$ eV) and on the TCR (-0.028 K^{-1}), and therefore in a decrement on the pixel performance.

In our work we have studied the electrical and optical properties of amorphous germanium-silicon (a-Ge$_x$Si$_y$:H) and amorphous germanium-silicon-boron (a-Ge$_x$Si$_y$B$_z$:H) thin films deposited by plasma (PECVD) (R. Ambrosio, 2004; A. Kosarev, 2006; A. Torres, 2008; M. Moreno, 2007, 2008, 2010). Intrinsic a-Ge$_x$Si$_y$:H has better performance characteristics than a-Si:H,B when is used as thermo-sensing element, since it has a high activation energy ($E_a = 0.37$ eV), a high TCR ($\alpha = -0.047$ K^{-1}), a moderated room temperature conductivity ($\sigma_{RT} \approx 6x10^{-5}$ $(\Omega cm)^{-1}$), and therefore a moderated pixel resistance ($R_{pixel} \approx 30x10^7\ \Omega$) (M. Moreno, 2007, 2008) when is used as thermo sensing element in microbolometers.

In the other hand a-Ge$_x$Si$_y$B$_z$:H has an improved room temperature conductivity ($\sigma_{RT} \approx 10^{-2}$-$10^{-3}$ $(\Omega cm)^{-1}$), and therefore a moderated pixel resistance ($R_{pixel} \approx 1$-$5x10^6\ \Omega$) (M. Moreno, 2007, 2008), but also it has a low activation energy ($E_a \approx 0.18 - 0.22$ eV) and low the TCR (-0.023 -0.028 K^{-1}).

In this chapter we present a summary on the study of a-Ge$_x$Si$_y$:H and a-Ge$_x$Si$_y$B$_z$:H thin films and their application as thermo-sensing element in microbolometers. We have fabricated, characterized and studied two devices configurations labeled as planar (the standard configuration used in commercial microbolometer arrays) and sandwich structures. The later shows several advantages when intrinsic materials are used as thermo-sensing element. Finally we studied the performance characteristics of the different device configurations and compared them with commercial devices and those reported on literature.

2. Principle of operation of un-cooled microbolometers

The operation of a microbolometer is based on the temperature rise of the thermo-sensing material by the absorption of the incident IR radiation. The change in temperature causes a change on its electrical resistance, which is measured by an external circuit. Microbolometers based on amorphous semiconductors have advantages over other types of thermal detectors, including microbolometers that use other kind of thermo-sensing materials. The advantages are mainly technological, since these microbolometers are fully compatible with silicon CMOS fabrication technology, there is no need of additional fabrication equipment in a IC production line. Are relatively of simple fabrication and can be processed at relatively low temperature by PECVD. The above make them ideal for a post-process fabrication over a CMOS read-out circuit.

Fig. 1 shows a scheme of one microbolometer (B. E. Cole, 1998); it is built on a membrane usually made of SiN$_x$. Over the membrane is deposited the thermo-sensing material and the IR absorber material. The membrane provides thermal isolation to the thermo-sensing film.

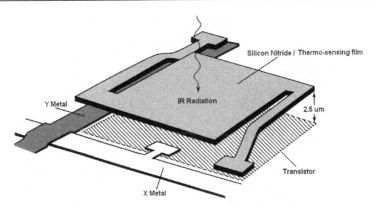

Fig. 1. Microbolometer scheme.

2.1 Thermal insulation

There are three mechanism of heat transfer that occurs in a thermal detector, they are conduction, convection and radiation. Conduction mechanisms occur when the heat flows from the thermo-sensing area along the supporting legs to the substrate. Conduction is critical when the pixels are very close, since the heat can flow from one pixel to a neighbor pixel. Convection occurs when the heat flows in the presence of a surrounding atmosphere, this mechanism is not very important if the detector is encapsulated in a vacuum package. Radiation mechanism is presented by the fact that the detector radiates to its surroundings and the surroundings radiate to it.

When the microbolometers are encapsulated in an evacuated package, with an IR transmitting window, convection and radiation mechanism are minimized. Thus the main loss of heat mechanism is conduction from the thermo-sensing material to the substrate through the supporting structure.

The supporting structure is a very important part of thermal detectors, it provides three functions, mechanical support, electrical conducting path and thermal conducting path. In order to avoid heat losses in microbolometers, it is necessary to improve the thermal insulation. In microbolometers there are two main thermal insulation configurations: single-level and two-level configurations.

Single level configuration consist in deposit a membrane over the silicon (Si) substrate and after that, open a hole in the Si substrate, employing bulk micromachining techniques. Bulk micromachining consumes area, since the Si substrate is etched with a side wall angel of 54.9 degrees. The electronic circuit (which forms part of the read out circuit) is fabricated next to the pixel, consuming area also. That result in a 20% fill factor.

The two-level configuration allows the fabrication of the electronics circuit in the substrate and after that, the fabrication of the microbolometer in a low temperature post process over the electronics, by using surface micromachining techniques. With this configuration is saved substrate area, achieving a fill factor of above 70%.

In order to fabricate thermal sensors in a post process, over a wafer surface, containing an IC circuit; it is necessary to use low temperatures during the fabrication process. By employing

Plasma Enhanced Chemical Vapor deposition (PECVD) it is possible to deposit thin films at relatively low temperatures (150 - 350 ºC).

2.2 Infrared absorber films

An absorber element is a very important part in un-cooled IR microbolometers, its role is based in the absorption of IR radiation and the transfer of heat to the thermo-sensing material. The main requirements of absorbing materials for un-cooled microbolometers are: A high absorbance coefficient in the range λ = 8 – 12 µm, simple fabrication and compatibility with the silicon CMOS technology.

The IR absorption can be improved employing a resonant micro-cavity, where the thermo-sensing film is separated from the substrate by a gap equivalent to one quarter of the wavelength at which it will be operating. A mirror (Al or Ti) is deposited over the substrate surface, under the thermo-sensing material. In this configuration the radiation that was not absorbed by the thermo-sensing film will resound inside the cavity and will be re-absorbed by the thermo-sensing element.

Terrestrial objects have temperatures around of 300K, with IR emission centered in 10 µm. Thus un-cooled microbolometers employed for detection of objects at room temperature, should have a gap from the substrate of 2.5 µm, for the fabrication of the resonant micro-cavity.

Several materials have been employed as absorbing films in microbolometers, which are deposited over the thermo-sensing film. Among the most employed are some metals, as black gold film (M. Hirota, 1998), which has a very high absorption coefficient of IR radiation (more than 90 %), however it is not a standard material in CMOS technology. SiN_x films are employed commonly as absorber films in microbolometers (A. Schaufelbühl, 2001, S. Sedky, 1998), since its absorption coefficient can be tuned by the deposition parameters and it is a standard material in CMOS technology.

2.3 Thermo-sensing films

The thermo-sensing material is perhaps the most important element in a microbolometer. The increment in temperature in the sensing material causes a change in some temperature-dependent parameter. In the case of a microbolometer that parameter is the resistance.

The thermo-sensing material should have a large temperature coefficient of resistance, TCR ($\alpha(T)$), which is defined by Eq. 1, where E_a is the activation energy, K is the Boltzman constant and T is temperature.

$$\alpha(T) = (1/R)[dR/dT] \approx E_a / KT^2 \tag{1}$$

A large TCR means that a small change in temperature in the sensing material will result in a large change in resistance. Eq. 1 shows that the TCR and E_a are directly related, thus a high E_a in the material is needed.

For un-cooled microbolometers vanadium oxide, VO_x, was the first thermo-sensing element employed (B. E. Cole, 1998), since it has a moderated TCR, $\alpha(T) \approx 0.021$ K^{-1}, however it is not a standard material in silicon CMOS technology. Some metals have been employed also, which are compatible with Si-CMOS technology, however they have low values of TCR (Pt, $\alpha(T)\approx 0.0015$ K^{-1}).

Hydrogenated amorphous silicon (a-Si:H) prepared by PECVD is very attractive to be used as thermo-sensing film in microbolometers, for room temperature operation (A. J. Syllaios, 2000). It is compatible with the IC technology, has a high activation energy, $E_a \approx 0.8 - 1$ eV and high value of TCR, $\alpha(T) \approx 0.1 - 0.13$ K^{-1}, however it also has a very high undesirable resistivity, which often cause a mismatch with the input impedance of the read-out circuits. In order to reduce the a-Si:H high resistance, boron doping has been employed. The B doped a-Si:H films have a significant reduction in its resistivity, however a reduction in E_a and TCR is obtained also, Ea ≈ 0.22 eV and TCR ≈ 0.028 K^{-1} (A. J. Syllaios, 2000).

Material	TCR (K^{-1})	E$_a$ (eV)	σ_{RT} (Ω cm)$^{-1}$	Reference
VO$_x$	0.021	0.16	2x10^{-1}	B. E. Cole, 1998
a-Si:H (PECVD)	0.1 - 0.13	0.8-1	~ 1x10^{-9}	A. J. Syllaios, 2000
a-Si:H,B (PECVD)	0.028	0.22	5x10^{-3}	A. J. Syllaios, 2000
a-Ge$_x$Si$_y$:H (PECVD)	0.043	0.34	1.6x10^{-6}	M. Moreno, 2008
Poly-SiGe	0.024	0.18	9x10^{-2}	S. Sedky, 1998
Ge$_x$Si$_{1-x}$O$_y$	0.042	0.32	2.6x10^{-2}	E. Iborra, 2002
YBaCuO	0.033	0.26	1x10^{-3}	J. Delerue, 2003

Table 1. Common materials employed as thermo-sensing films in microbolometers.

In our work (R. Ambrosio, 2004; A. Kosarev, 2006; A. Torres, 2008; M. Moreno, 2007, 2008, 2010), amorphous germanium-silicon, a-Ge$_x$Si$_y$:H, deposited by PECVD has been studied as thermo-sensing films in un-cooled microbolometer, obtaining high activation energy, E_a= 0.34 eV, consequently a high value of TCR = 0.043 K^{-1} and improved but still high resistivity.

Table 1 shows the most common materials employed as thermo-sensing films in microbolometers. As can be seen in the table, there are available several materials which can be used as thermo-sensing films. Intrinsic amorphous silicon, a-Si:H and a-Ge$_x$Si$_y$:H, show the largest TCR values and are fully compatible with the silicon CMOS technology, however they have also the smallest values of room temperature conductivity, σ_{RT}.

3. Main figures of merit of un-cooled microbolometers

In this section the different figures of merit of a microbolometer, as thermal characteristics, responsivity and detectivity are presented. The different types of noise in microbolometer are described also.

3.1 Thermal capacitance, C$_{th}$, thermal conductance, G$_{th}$ and thermal response time, τ_{th}

A simple representation of a microbolometer is shown in Fig. 2, the detector has a thermal capacitance, C$_{th}$, and it is coupled to the substrate which is a heat sink, by a thermal conductance, G$_{th}$.

When the detector receives modulated IR radiation, the rise in temperature is found by solving the balance equation, Eq. 2; where C_{th} (expressed in JK^{-1}) is the thermal capacitance of the supporting membrane containing the thermo sensing film, while G_{th} (expressed in WK^{-1}) is the thermal conductance of the legs, which is considered the main heat loss

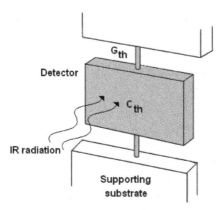

Fig. 2. Microbolometer representation.

mechanism. ΔT is the temperature difference of the hot and reference junctions. A_{cell} is the detector area, β is the fill factor, which is the radio of the thermo-sensing film area to the total cell area, η is the optical absorption coefficient, defined as the fraction of the radiant power falling on the thermo-sensing area, which is absorbed by that area. P_o is the intensity of the IR modulated radiation, ω is the angular modulation frequency and t is time [2.6].

$$C_{th}d(\Delta T)/dt + G_{th}(\Delta T) = \eta\beta A_{cell} P_0 \exp(j\omega t) \qquad (2)$$

The solution of the balance equation is shown in Eq. 3:

$$\Delta T = \eta\beta A_{cell} P_0 / G_{th}(1+\omega^2\tau_{th}^2)^{1/2} \qquad (3)$$

Where, τ_{th} (expressed in seconds) is the thermal response time of the microbolometer, it is defined by Eq. 4, which establishes a relation between τ_{th}, C_{th} and G_{th}. Typical values of thermal time constant are in the range of milliseconds, which are much longer than the typical time of photon detectors.

$$\tau_{th} = C_{th} / G_{th} \qquad (4)$$

For unmodulated radiation Eq. 3 can be reduced to:

$$\Delta T(\omega = 0) = \eta\beta A_{cell} P_0 / G_{th} \qquad (5)$$

Eq. 5 shows that the increment of temperature, ΔT, in the detector is inversely proportional to the thermal conductance G_{th} of its legs. In order to achieve a high performance microbolometers ΔT should be as high as possible and therefore G_{th} as small as possible, which can be done by making very thin the detector legs.

3.2 Responsivity

Responsivity, R, is defined as the ratio of the pixel output signal to the incident radiant power (in Watts) falling on the pixel (P. W. Kruse, 2001). The output signal is an electrical signal that can be voltage or current, thus R can be expressed in Volts/Watts (voltage responsivity, R_u) or Amps/Watts (current responsivity, R_I). In order to obtain R, we can use the simplest model, where it is assumed that there is no heating due the electrical bias in the detector (Joulean heating), and also it is assumed a constant electrical bias to the detector.

When the microbolometer is current biased, the output signal is voltage, V_s, given by Eq. 6, where I_b is the bias current, R_{cell} is the electrical resistance of the microbolometer, α is the TCR, described by Eq. 1 and ΔT is the increment of temperature in the detector, obtained in Eq. 5.

$$V_s = I_b R_{cell} \alpha \, \Delta T \tag{6}$$

Voltage responsivity, R_v, is obtained by combining equations 3 and 6, and dividing by $P_o A_{cell}$, which is the incident radiant power, the result is shown in Eq. 7.

$$R_v = \eta \, \beta \, I_b \alpha R_{cell} \, / \, G_{th}\left(1 + \omega^2 \tau_{th}^{\,2}\right)^{1/2} \tag{7}$$

$$R_v = \eta \, \beta \, I_b \alpha R_{cell} \, / \, G_{th} \tag{8}$$

For unmodulated radiation, $\omega = 0$, Eq. 7 is simplified in Eq. 8, which is the DC responsivity. When the microbolometer is voltage biased equations 7 and 8 are transformed to Eq. 9 and Eq. 10 respectively, where R_I is current responsivity.

$$R_I = \eta \, \beta \, V_b \alpha \, / \, G_{th} R_{cell}\left(1 + \omega^2 \tau_{th}^{\,2}\right)^{1/2} \tag{9}$$

$$R_I = \eta \, \beta \, V_b \alpha \, / \, G_{th} R_{cell} \tag{10}$$

3.3 Noise in microbolometers

There are four main sources of noise in microbolometers (P. W. Kruse, 2001), which are Johnson noise, 1/f noise, temperature fluctuation noise and background fluctuation noise, these noise types are uncorrelated and are described in the following subsections.

3.3.1 Johnson noise

The Johnson noise component, V_j, is described by Eq. 11, where k is the Boltzmann constant, T_{cell} is the bolometer temperature, R_{cell} is the bolometer resistance and Δf is the bandwidth of the integration time.

$$V_j = \left(4k \, T_{cell} R_{cell} \Delta f\right)^{1/2} \tag{11}$$

3.3.2 1/f noise

The 1/f noise is characterized by a spectrum that depends inversely on frequency and is described by Eq. 12, where V is the product of the bias current - I_b and the electrical resistance of the microbolometer - R_{cell}, f is the frequency at which the noise is measured and n is the 1/f noise parameter, which depend on the material detector.

$$V_{1/f} = \left(V^2 n \ / \ f\right)^{1/2} \tag{12}$$

1/f noise is the dominant noise at low frequencies and falls below the Johnson noise at higher frequencies, the transition is commonly called the "knee".

3.3.3 Temperature fluctuation noise

A thermal detector which is in contact with its environment (by conduction and radiation), exhibits random fluctuations in temperature, since the interchange of heat with its surrounding has a statistical nature; this is known as temperature fluctuation noise. The mean square temperature fluctuation noise voltage is given by Eq. 13 (P. W. Kruse, 2001).

$$V_{TF}^2 = \left(4kT_{cell}^2 \Delta f \ / \ G\left(1 + \omega^2 \tau_{th}^2\right)^{1/2}\right) V^2 \alpha^2 \tag{13}$$

3.3.4 Background fluctuation noise

When the heat exchange by conduction between the detector and its surroundings is negligible, in comparison with the radiation exchange, the temperature fluctuation noise will be identified as background fluctuation noise.

The mean square background fluctuation noise is given by Eq. 14, where T_{cell} is the detector temperature and T_B is the background temperature.

$$V_{BF}^2 = 8 \, A_{cell} \eta \ \sigma \, k \left(T_{cell}^5 + T_B^5\right) R_{cell}^2 \tag{14}$$

The total noise voltage is obtained by adding the 4 noise contributions as is shown in Eq. 15.

$$V_N^2 = Vj^2 + V_{1/f}^2 + V_{TF}^2 + V_{BF}^2 \tag{15}$$

3.4 Detectivity

Detectivity, D^* (expressed in $cmHz^{1/2}Watt^{-1}$), is a figure of merit for all types of detectors, it is defined as the pixel output signal to noise ratio per unit of incident radiant power falling on the detector, measured in a 1 Hz bandwidth. In other words, D^* is the normalized signal to noise ratio in the detector and is shown in Eq. 16.

$$D^* = \left(R_V \left(A_{cell} \Delta_f\right)^{1/2}\right) \ / \ V_N \tag{16}$$

In Eq. 16 R_v is the voltage responsivity, A_{cell} is the detector area, Δf is the frequency bandwidth and V_N is the contribution of the four noises. It is clear that in order to achieve a high D^* the responsivity should be as high as possible and the noise as small as possible.

The fundamental limit to sensitivity of any thermal detector is set by random fluctuations in the temperature of the detector due to fluctuations in the radiant power exchange between the detector and its surroundings. The highest possible value of D^* of a thermal detector operated at room temperature is $D^* = 1.98 \times 10^{10} \ cmHz^{1/2}W^{-1}$(A. Rogalski, 2003).

4. Amorphous germanium-silicon (a-Ge$_x$Si$_y$:H) and germanium-silicon-boron alloys (a-Ge$_x$Si$_y$B$_z$:H)

Intrinsic amorphous silicon (a-Si:H) prepared by PECVD is a very attractive material to be used in microbolometers as thermo-sensing film. It has a high activation energy, $E_a \approx 0.8 - 1$ eV and high value of temperature coefficient of resistance, TCR, $\alpha(T) \approx 0.1 - 0.13$ K^{-1}, however it also has a high undesirable resistivity.

Amorphous germanium-silicon (a-Ge$_x$Si$_y$:H) films deposited by PECVD have been studied as thermo-sensing film in microbolometers (R. Ambrosio, 2004; A. Kosarev, 2006; A. Torres, 2008; M. Moreno, 2007, 2008, 2010), due its high activation energy and consequently high TCR, and its relatively high room temperature conductivity, σ_{RT}, in comparison with a-Si:H films. In this section is presented a description of the deposition by PECVD of intrinsic amorphous germanium-silicon (a-Ge$_x$Si$_y$:H) and amorphous germanium-silicon-boron (a-Ge$_x$Si$_y$B$_z$:H) thin films, and its electrical and compositional characterization.

4.1 Films preparation for characterization

An intrinsic film (a-Ge$_x$Si$_y$:H) was deposited in a capacitive discharge low frequency (LF) PECVD reactor at frequency f = 110 KHz, substrate temperature T$_s$ = 300 ℃, pressure P = 0.6 Torr and RF power W = 350 W, with a gas mixture of SiH$_4$, GeH$_4$ and H$_2$ and gas flow rates of Q$_{SiH4}$=25 sccm, Q$_{GeH4}$ =25 sccm and Q$_{H2}$=1000 sccm respectively. This result in a Ge gas content X$_g$ = 0.5. The film was labeled as process A.

The a-Ge$_x$Si$_y$B$_z$:H films were also deposited in a capacitive discharge low frequency (LF) PECVD reactor at frequency f = 110 KHz, substrate temperature T$_s$ = 300 ℃, pressure P = 0.6 Torr and RF power W = 350 W. Three sets of films were deposited from SiH$_4$ (100%), GeH$_4$ (100%) and B$_2$H$_6$ (1% on H$_2$) gas mixture, with a fixed SiH$_4$ and B$_2$H$_6$ gas flow rates: Q$_{SiH4}$=50sccm and Q$_{B2H6}$=500 sccm, respectively, while the GeH$_4$ gas flow was set at the following values: Q$_{GeH4}$ =25, 50 and 75 sccm.

The late resulted in a Ge gas content X$_g$= 0.3, 0.45, 0.55 and a B gas content Z$_g$= 0.11, 0.09, 0.07 in the samples labeled as process number B, C and D, respectively. Table 2 shows the deposition parameters for the 4 thermo-sensing films.

	Process A (intrinsic)	Process B	Process C	Process D
Gases flow rates (sccm)	SiH$_4$(100%): 25 GeH$_4$(100%):25 H$_2$: 1000	SiH$_4$ (100%): 50 GeH$_4$(100%): 25 B$_2$H$_6$ (1%): 500	SiH$_4$ (100%): 50 GeH$_4$(100%): 50 B$_2$H$_6$ (1%): 500	SiH$_4$ (100%): 50 GeH$_4$(100%): 75 B$_2$H$_6$ (1%): 500
Dilution ratio (%): $\dfrac{H_2}{GeH_4+SiH_4+B_2H_6}$	20	6.2	4.7	3.8
Ge contentin gas mixture (%)	50	30	45	55
B contentin gas mixture (%)	----	11	9	7
Temperature (℃)	300℃	300℃	300℃	300℃
Pressure (Torr.)	0.6	0.6	0.6	0.6
Frequency (Khz.)	110	110	110	110
Power (W)	300	300	300	300

Table 2. Deposition parameters of a-Ge$_x$Si$_y$:H and a-Ge$_x$Si$_y$B$_z$:H films.

Since those films are studied for applications as thermo-sensing films for microbolometers, we measured the film electrical properties after patterning them with photolithography in one cell of dimensions $70 \times 66 \ \mu m^2$.

Assuming that stress arisen in the film deposited over a SiN_x micro-bridge could have an effect on the film conductivity, we also studied the films deposited on a micro-bridge. For that purpose, we prepared three different kinds of samples for each type of the four thermo-sensing films (three boron alloys with different Ge content and the intrinsic film). The films were prepared as is shown in Fig. 3.

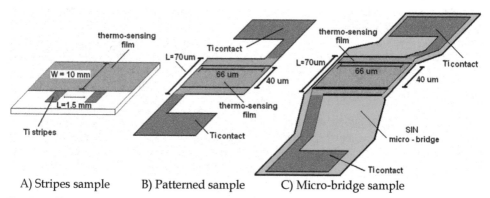

A) Stripes sample   B) Patterned sample   C) Micro-bridge sample

Fig. 3. Different thermo-sensing films samples. A) Stripes, B) pattern and C) Micro-bridge.

4.2 Temperature dependence of conductivity and TCR in a-Ge$_x$Si$_y$:H and a-Ge$_x$Si$_y$B$_z$:H films

We performed measurements of temperature dependence of conductivity $\sigma(T)$ in the a-Ge$_x$Si$_y$:H and a-Ge$_x$Si$_y$B$_z$:H thermo-sensing films in the range of T= 300–400 K. The measurements were performed in a vacuum chamber at a pressure P≈20 mTorrs. A temperature controller (model K-20, MMR Inst.) for the temperature measurement control and an electrometer (model 6517-A, Keithley Inst.) for the current measurements were employed. These measurements allowed us to obtain the $\sigma(T)$ temperature dependence and then to determine the E_a, the TCR and the room temperature conductivity, σ_{RT}.

The conductivity temperature dependence can be well described by $\sigma(T)=\sigma_0 \exp(-E_a/kT)$, where σ_0 is the prefactor, E_a is the activation energy, k is the Boltzmann constant and T is the temperature. Fig. 4 shows $\sigma(T)$ curves for four different thermo-sensing films (three boron alloys with different Ge gas content, Ge$_x$ = 0.3, 0.45, 0.55 and the intrinsic film with Ge$_x$ = 0.5), fabricated in three different sample configurations (stripes, patterns and micro-bridges).

From $\sigma(T)$ measurements with temperature in the thermo-sensing films, we found that the boron alloys (a-Ge$_x$Si$_y$B$_z$:H) have a significantly larger conductivity (by about 2-3 orders of magnitude) in comparison with that of the intrinsic film (a-Ge$_x$Si$_y$:H). We observed that an increment in the Ge content in gas phase in the boron alloys results in an increase of the room temperature conductivity, from σ_{RT} = 2.8 x10^{-3} (Ωcm)$^{-1}$ (for Ge$_x$ = 0.3) to σ_{RT} = 1 x10^{-2} (Ωcm)$^{-1}$ (for Ge$_x$ = 0.45) and σ_{RT} = 2.5 x10^{-2} (Ωcm)$^{-1}$ (for Ge$_x$ = 0.55), while for the intrinsic film the room temperature conductivity is σ_{RT} = 6 x10^{-5} (Ωcm)$^{-1}$ (for Ge$_x$ = 0.5).

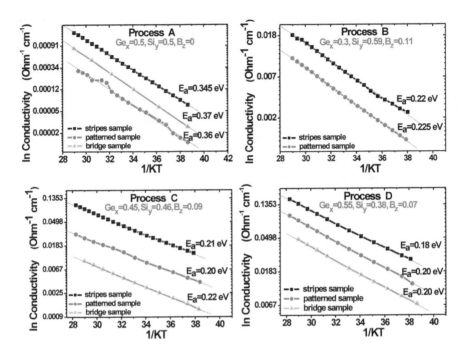

Fig. 4. Conductivity dependence with temperature for the different thermo-sensing films (process: A, B, C and D).

The increment in the σ is accompanied with a reduction in the E_a. We obtained an $E_a= 0.22$ eV (for $Ge_x = 0.3$), $E_a= 0.21$ eV (for $Ge_x = 0.45$) and $E_a= 0.18$ eV (for $Ge_x = 0.55$), while in the intrinsic film is $E_a= 0.345$ eV (for $Ge_x = 0.5$). E_a as a function of Ge_x is shown in Fig. 5 A).

The reduction in the thermo-sensing films dimensions, from the stripes samples ($10x1.5$ mm²) to the patterned samples (70×66 μm²), has no significant effect on E_a, however it has on the σ_{RT}. We observed a reduction of above 50-80 % of the σ_{RT} value in the patterned samples in comparison with that of the stripes samples.

Practically no change in E_a of the thermo-sensing films deposited over a SiN_x micro-bridge was observed, in comparison with that of the stripes and patterned samples; however the micro-bridge samples showed a larger reduction in the σ_{RT} values, of 60-90 %. The dependence of σ_{RT} with the Ge_x content and the sample structure are shown in Fig. 5 B), while the deposition rate dependence of Ge_x content in the thermo-sensing films is shown in Fig. 5 C). Table 3 show a comparison of E_a, TCR, σ_{RT} and σ_0 in stripes, patterned and micro-bridges samples for the different thermo-sensing films.

The micro-bridges samples have the largest reduction of conductivity, and it could be explained by the stress arisen in the SiN_x micro-bridge, affecting the thermo-sensing film electrical conductivity. The deposition rate in the boron alloys is around 2 -3 times larger than that of the intrinsic film. Boron incorporation during the thermo-sensing deposition, enhance the deposition rate as is shown in Fig. 5 C).

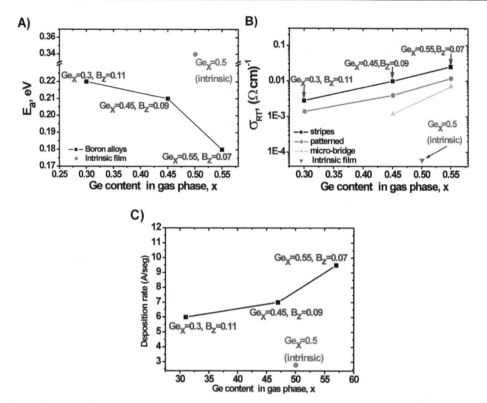

Fig. 5. Characterization of a-Ge$_x$Si$_y$:H and a-Ge$_x$Si$_y$B$_z$:H films. A) E$_a$ as function of Ge gas content (Ge$_x$). B) Conductivity as a function of Ge gas content. C) Deposition rate as a function of Ge gas content.

It is important to point out that, doping on amorphous semiconductors reduce E$_a$ and increases the films conductivity (σ_{RT}). In Fig 5A) for reference, is shown an intrinsic a-Ge$_x$Si$_y$:H film produced with a gas content of Ge$_x$=50% and Si$_y$=50%, which has a E$_a$ of 0.34 eV. This is the largest value for a-Ge$_x$Si$_y$:H films (doped or un-doped, using a gas content of Ge$_x$=50% and Si$_y$=50%).

When boron is introduced in the film deposition, E$_a$ is reduced and the conductivity is increased in the films. In fig. 5 A) is shown that E$_a$ is reduced to values in the range of 0.18 - 0.22 eV, while the conductivity is increased in more than one order of magnitude. Also it is important to notice that a-Ge$_x$Si$_y$:H films have an intermediate E$_a$ value, between a-Si:H and a-Ge:H. Intrinsic a-Si:H has E$_a$ values close to 1 eV, while a-Ge:H have Ea values of above 0.3 eV. Thus, varying the Ge (and Si) gas contents in the a-Ge$_x$Si$_y$:H films, it is possible to modify E$_a$ (an also the conductivity) on intrinsic films.

Larger Ge$_x$ content in the films will reduce the value of E$_a$. In fig 5 A), we observe a decrement on E$_a$ of the a-Ge$_x$Si$_y$B$_z$:H films, not just because the B$_z$ gas content (which in fact decreases), but because the Ge content (which increases). In fig 5 A) is shown that for a-Ge$_x$Si$_y$B$_z$:H films, the Ge$_x$ gas content vary between 0.3 and 0.55, while the B$_z$ gas content vary just between 0.07 and 0.11. Thus the effect of the variation of the Ge gas content on E_a is dominant in the a-Ge$_x$Si$_y$B$_z$:H films .

4.3 Composition of the a-Ge$_x$Si$_y$:H and a-Ge$_x$Si$_y$B$_z$:H films

In Figure 5 the showed results are related to the Ge$_x$, Si$_y$, and B$_z$ gas contents, not to solid contents. There exists a significant difference between the gas content used for the films deposition, and the solid content in the films produced. The composition in solid phase of the different films (three boron alloys with different Ge gas content and the intrinsic film) was characterized by secondary ion mass spectroscopy (SIMS). The samples used for SIMS characterization were the stripes samples described in section 4.1. Fig. 6 shows the SIMS profiles obtained.

From SIMS profiles we calculated the solid composition in the thermo-sensing films. For the film with gas content: Ge$_x$=0.3 and B$_z$=0.11 (process B), we observed an increase in the solid content: Ge$_x$=0.59 and B$_z$=0.32 respectively. For the film with Ge$_x$=0.45 and B$_z$=0.09 (process C), we observed Ge$_x$=0.67 and B$_z$=0.26, respectively. For the film with Ge$_x$=0.55 and Bz=0.07 (process D), we observed Ge$_x$=0.71 and B$_z$=0.23, respectively. These results suggested a strong preferential B and Ge incorporation from gas phase during the film deposition process. The B$_z$ solid content demonstrated values about 3 times larger than the content in gas phase B$_z$, while the Ge$_x$ solid content increased by a factor of 1.3-2 from the Ge$_x$ gas content. Those results are shown in Table 4.

		Thermo-sensing films			
		Process A	Process B	Process C	Process D
Film Thickness (µm)		0.5	0.36	0.42	0.51
Deposition rate (A/s)		2.8	6	7	9.5
Stripes samples	E_a (eV)	0.345	0.22	0.21	0.18
	TCR (K^{-1})	-0.044	-0.028	-0.027	-0.023
	σ_{RT} (Ωcm)$^{-1}$	6x10^{-5}	2.8x10^{-3}	1x10^{-2}	2.5x10^{-2}
	σ_0 (Ωcm)$^{-1}$	34.85	12.02	36.46	24.55
Patterned samples	E_a (eV)	0.36	0.225	0.20	0.20
	TCR (K^{-1})	-0.046	-0.029	-0.025	-0.025
	σ_{RT} (Ωcm)$^{-1}$	1.08x10^{-5}	1.4x10^{-3}	4x10^{-3}	1.2x10^{-2}
	σ_0 (Ωcm)$^{-1}$	11.13	7.27	8.23	28.26
Micro-bridge samples	E_a (eV)	0.37	Not available	0.22	0.20
	TCR (K^{-1})	-0.047		-0.028	-0.025
	σ_{RT} (Ωcm)$^{-1}$	2.2x10^{-5}		1.2x10^{-3}	7x10^{-3}
	σ_0 (Ωcm)$^{-1}$	32.8		5.94	15.58

Table 3. Comparison of E_a, TCR, σ_{RT} and σ_0 in stripes, patterned and micro-bridges samples for the different thermo-sensing films.

In Fig. 6 the intrinsic film has a Boron content of 10^{18} cm^{-3} (which represents a B solid content of 2x10^{-3} %) as is shown in table 4. The reason of the above is the fact that all the films were deposited on the same chamber. Even though, the chamber was extensively cleaned and coated with a SiN$_x$ film before the intrinsic film deposition, Boron impurities remained in the chamber walls, which were re-deposited in the intrinsic films.

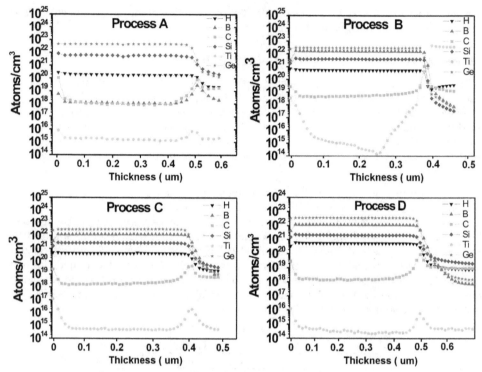

Fig. 6. SIMS profiles of a-Ge$_x$Si$_y$:H and a-Ge$_x$Si$_y$B$_z$:H thermo-sensing films.

		Thermo-sensing films			
		Process A	Process B	Process C	Process D
Gas content	Ge$_x$ (%)	0.5	0.3	0.45	0.55
	Si$_y$ (%)	0.5	0.59	0.46	0.38
	B$_z$ (%)	0	0.11	0.09	0.07
Solid content obtained from SIMS	Ge$_x$ (%)	0.888	0.59	0.67	0.71
	Si$_y$ (%)	0.110	0.078	0.05	0.04
	B$_z$ (%)	2.0×10^{-3}	0.32	0.26	0.23

Table 4. Gas content and solid content obtained by SIMS for the thermo-sensing films.

5. Microbolometer configurations and fabrication process flow

In this section we show a comparative study of the performance characteristics of three configurations of un-cooled microbolometers based on amorphous germanium thin films: a) Planar structure with intrinsic amorphous germanium-silicon a-Ge$_x$Si$_y$:H thermo-sensing film. In this configuration the metal electrodes are placed under the thermo-sensing film (Fig. 7 A); b) Planar structure with amorphous germanium-boron-silicon alloy a-Ge$_x$B$_y$Si$_z$:H thermo-sensing film (Fig. 7 B) and c) Sandwich structure with intrinsic a-Ge$_x$Si$_y$:H thermo-sensing film, this configuration consists of metal electrodes which sandwich the thermo-sensing film (Fig. 7 C). Fig. 7 D) shows a picture of one device fabricated.

The fabrication process of the planar structure microbolometer with the a-Ge$_x$Si$_y$:H thermo-sensing film is as follows. A 0.2 µm-thick SiO$_2$ layer is deposited by CVD on a c-Si wafer and a 2.5 µm-thick sacrificial aluminum layer is deposited by e-beam evaporation and patterned. A 0.8 µm-thick SiN$_x$ film is then deposited at low temperature (350 °C) by low frequency PECVD over the aluminum sacrificial film. The SiN$_x$ film is patterned by reactive ion etching (RIE) in order to form a SiN$_x$ bridge. A 0.2 µm-thick titanium contacts are deposited by e-beam evaporation over the SiN$_x$ bridge and a 0.5 µm-thick thermo-sensing a-Ge$_x$Si$_y$:H film is deposited over the Ti contacts by low frequency LF PECVD technique at a rf frequency f=110 kHz, temperature T=300 °C, power W=350 W and pressure P=0.6 Torr. The a-Ge$_x$Si$_y$:H film is deposited from a SiH$_4$ + GeH$_4$ + H$_2$ mixture with gas flows: Q$_{SiH4}$=25sccm, Q$_{GeH4}$ =25 sccm, Q$_{H2}$=1000 sccm. This results in a Ge content in solid phase Y=0.88 and a Si content in solid phase Y=0.11. The thermo-sensing film is covered with a 0.2 µm-thick absorbing SiN$_x$ film deposited by PECVD and finally the aluminum sacrificial layer is removed with wet etching.

The planar structure microbolometer with the boron alloy (a-Ge$_x$B$_y$Si$_y$:H) thermo-sensing film is fabricated as the previous one, with difference in the thermo-sensing film deposition parameters. The boron alloy film is deposited from a SiH$_4$ + GeH$_4$ + B$_2$H$_6$ + H$_2$ mixture with the following gas flows: Q$_{SiH4}$=50sccm, Q$_{GeH4}$ =50 sccm, Q$_{B2H6}$=5 sccm and Q$_{H2}$=500 sccm. This results in a Ge content in solid phase Ge$_x$=0.67 and B content in solid phase B$_y$=0.26. Those values were obtained by SIMS measurements.

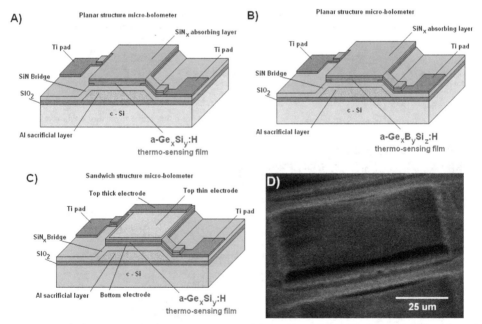

Fig. 7. Microbolometers: A) Planar wit intrinsic film a-Ge$_x$Si$_y$:H, B) Planar with boron doped film a-Ge$_x$B$_y$Si$_z$:H, C) Sandwich with intrinsic film a-Ge$_x$Si$_y$:H, D) A device fabricated.

The sandwich structure microbolometer with the a-Ge$_x$Si$_y$:H film is fabricated in the same way as the planar microbolometer with some differences, due to the placing of metals as bottom and top electrodes. In this structure the electrodes sandwich the thermo-sensing film. The bottom Ti electrode is 0.2 μm-thick and is deposited before the thermo-sensing film. Then the a-Ge$_x$Si$_y$:H film is deposited and it is covered with a top thin electrode (10 nm) forming a sandwich structure. The active area of the thermo-sensing layer in the three configurations studied is A$_b$=70x66μm². Fig. 8 shows the fabrication process of the microbolometer structures.

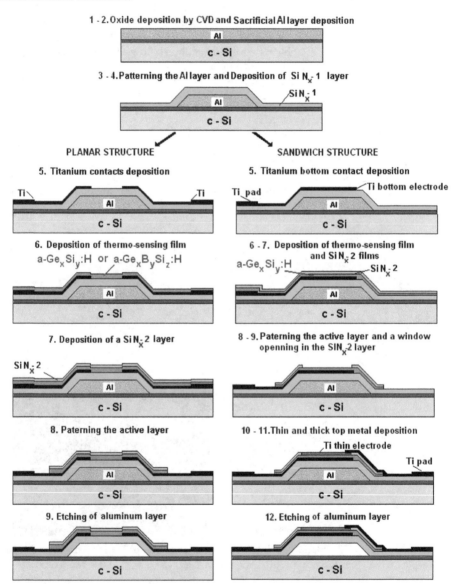

Fig. 8. Planar and sandwich microbolometers fabrication process flow.

6. Microbolometers electrical characterization

A microbolometer is a resistor sensitive to temperature change, its operation is based on the temperature increase of the thermo-sensing film by the absorption of the incident IR radiation. The change in temperature causes a change on its electrical resistance, which is measured by an external circuit.

In this section we present a comparative study of 3 configurations of un-cooled microbolometers based on amorphous silicon-germanium thin films deposited by plasma.

6.1 I(U) measurements in dark and under Infrared Radiation (IR)

In this section is described the procedure performed in order to obtain the current voltage I(U) characteristics of the microbolometers, from this measurement it is possible to determine the microbolometer electrical resistance and responsivity.

The current-voltage characteristics I(U) and current noise spectral density (NSD) have been measured in the devices in order to compare the performance characteristics, such as responsivity and detectivity in the 3 configurations of microbolometers:

a. Planar structure with an intrinsic germanium-silicon (a-Ge$_x$Si$_y$:H, Ge$_x$=0.5) thermo-sensing film (process A of section 4).
b. Planar structure with a germanium-silicon-boron alloy (a-Ge$_x$Si$_y$B$_z$:H, Ge$_x$=0.45, B$_z$=0.09) thermo-sensing film (process C of section 4).
c. Sandwich structure with an intrinsic (a-Ge$_x$Si$_y$:H, Ge$_x$=0.5) thermo-sensing film (process A of section 4).

The samples were placed in a vacuum chamber at pressure P≈20 mTorr, at room temperature and illuminated through a zinc selenide window (ZnSe). The window has a 70% transmission in the range of λ=0.6 – 20 μm. The source of IR light is a SiC globar source, which provides intensity I_0=5.3x10^{-2} W/cm^2 in the range of λ=1 – 20 μm. The current was measured with an electrometer ("Keithley"- 6517-A) controlled by a PC in dark and under IR illumination.

Fig. 9 A) shows the current-voltage I(U) characteristics in dark and under IR illumination for the planar configuration with a-Ge$_x$Si$_y$:H thermo-sensing film (process A, section 4); Fig. 9 B) shows these characteristics for the planar configuration with a-Ge$_x$Si$_y$B$_z$:H thermo-sensing film (process C, section 4); and Fig. 9 C) shows the same characteristics for the sandwich configuration with a-Ge$_x$Si$_y$:H thermo-sensing film (process A, section 4).

In those figures we can see the increment in current due to IR illumination, $\Delta I = I_{IR} - I_{Dark}$, where I_{IR} is the current under IR radiation and I_{Dark} is the current in dark. The planar configuration with the a-Ge$_x$Si$_y$:H (Ge$_x$=0.5) film has a ΔI = 5.4 nA (at bias voltage U=7 V); the planar configuration with the a-Ge$_x$Si$_y$B$_z$:H (Ge$_x$=0.45, B$_z$=0.09) film has a ΔI = 65 nA (at bias voltage U=7 V); and the sandwich configuration with the a-Ge$_x$Si$_y$:H (Ge$_x$=0.5) film has a ΔI = 35 μA (at bias voltage U=4 V). The inset in those figures show the Log I(Log U) characteristics, where we can see their linear behavior. The gamma (γ) constant indicates the slope of the curves.

6.2 Current and voltage responsivity

The current responsivity, R_I, is described by Eq. 17, where ΔI is the increment in current ($\Delta I = I_{IR} - I_{Dark}$) and $P_{incident}$ is the IR incident power in the device surface. $P_{incident}$ is described by Eq. 18 and is the product of the cell area, A_{cell} and the IR source intensity, I_0.

$$R_I = \Delta I \ / \ P_{incident} \qquad (17)$$

$$P_{incident} = A_{cell} I_0 \qquad (18)$$

The intensity of the IR source is $I_0 = 0.053$ Wcm^{-2}, while the cell area is $A_{cell} = (70 \times 10^{-4})(66 \times 10^{-4})$ cm^2 = 4.6×10^{-5} cm^2. Therefore the IR incident power in the device surface is $P_{incident} = 2.475 \times 10^{-6}$ W. The planar microbolometer with a-Ge$_x$Si$_y$:H (Ge$_x$=0.5) film has a $R_I = 2 \times 10^{-3}$ A/W (at U=7 V); the planar microbolometer with a-Ge$_x$Si$_y$B$_z$:H (Ge$_x$=0.45, B$_z$=0.09) film has a $R_I = 3 \times 10^{-2}$ A/W (at U=7 V); and the sandwich microbolometer with a-Ge$_x$Si$_y$:H (Ge$_x$=0.5) film has a $R_I = 14$ A/W (at U=4 V).

Table 5 shows the ΔI and R_I values obtained in the configurations. Fig. 10 A) shows the voltage dependence of R_I for the planar microbolometer with a-Ge$_x$Si$_y$:H (Ge$_x$=0.5) film; Fig. 10 B) shows the voltage dependence of R_I for the planar microbolometer with a-Ge$_x$Si$_y$B$_z$:H (Ge$_x$=0.45, B$_z$=0.09) film; and Fig. 10 C) shows the voltage dependence of R_I for the sandwich microbolometer with a-Ge$_x$Si$_y$:H (Ge$_x$=0.5) film. The insert in those figures show a relative current responsivity. Relative current responsivity is the ratio between the increment of current from dark to IR condition, ΔR_I, and the microbolometer resistance R_{cell}.

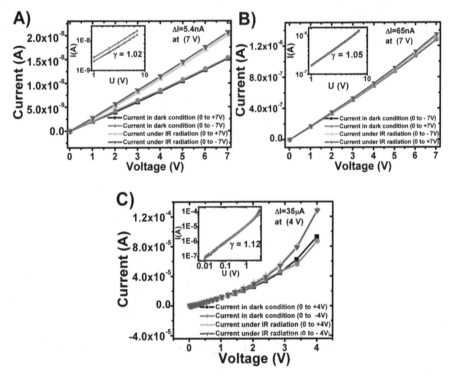

Fig. 9. I(U) characteristics of 3 microbolometers: A) planar with a-Ge$_x$Si$_y$:H (Ge$_x$=0.5). B) planar with a-Ge$_x$Si$_y$B$_z$:H (Ge$_x$=0.45, B$_z$=0.09). C) sandwich with a-Ge$_x$Si$_y$:H (Ge$_x$=0.5). The inset in those figures show the Log I(Log U) characteristics, where we can see their linear behavior. The gamma (γ) constant indicates the slope of the curves.

The planar and sandwich structures with the intrinsic film show larger values of relative current responsivity. The voltage responsivity, R_U, was calculated from the experimental I(U) points, the increment in voltage from dark condition to IR condition was obtained from a fixed current. Fig. 11 A) shows a $\Delta U = 1.8$ V extracted from a fixed current I = 1.5×10^{-8} A, in the planar structure microbolometer with the a-Ge$_x$Si$_y$:H (Ge$_x$=0.5) film. Fig. 11 B) shows a $\Delta U = 0.3$ V extracted from a fixed current I = 1.35×10^{-6} A, in the planar structure microbolometer with the a-Ge$_x$Si$_y$B$_z$:H (Ge$_x$=0.45, B$_z$=0.09) film. Fig. 11 C) shows a $\Delta U = 0.54$ V extracted from a fixed current I = 1.16×10^{-4} A, in the sandwich structure microbolometer with the a-Ge$_x$Si$_y$:H (Ge$_x$=0.5) film.

	Planar structureWith a-Ge$_x$Si$_y$:H	Planar structureWith a-Ge$_x$Si$_y$B$_z$:H	Sandwich structureWith a-Ge$_x$Si$_y$:H
Film process	A	C	A
$\Delta I = I_{IR} - I_{dark}$ (A)	5.4 x10^{-9} (at U=7 V)	65x10^{-9} (at U=7 V)	35x10^{-6} (at U=4 V)
Current responsivity (AW^{-1})	2x10^{-3}	3x10^{-2}	14
$\Delta U = U_{IR} - U_{dark}$ (V)	1.8 (at I=1.5x10^{-8})	0.3 (at I=1.35x10^{-6})	0.54 (at I=1.16x10^{-4})
Voltage Responsivity (VW^{-1})	7.2x10^5	1.2x10^5	2.2x10^5

Table 5. Current and voltage responsivity values for 3 microbolometer configurations.

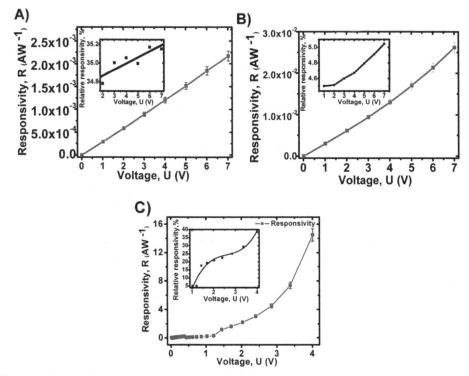

Fig. 10. Voltage dependence of R_I of 3 microbolometers: A) planar with a-Ge$_x$Si$_y$:H (Ge$_x$=0.5). B) planar with a-Ge$_x$Si$_y$B$_z$:H (Ge$_x$=0.45, B$_z$=0.09). C) sandwich with a-Ge$_x$Si$_y$:H (Ge$_x$=0.5).

The planar microbolometer with a-Ge$_x$Si$_y$:H (Ge$_x$=0.5) film has a R$_U$=7.2x10^5 V/W (at I=1.5x10^{-8} A); the planar microbolometer with a-Ge$_x$Si$_y$B$_z$:H (Ge$_x$=0.45, B$_z$=0.09) film has a R$_U$= 1.2x10^5 V/W (at I=1.4x10^{-6} A); and the sandwich microbolometer with a-Ge$_x$Si$_y$:H (Ge$_x$=0.5) film has a R$_U$= 2.2x10^5 V/W (at I=1.16x10^{-4} A). Table 5 shows the ΔU and R$_U$ values obtained from the different microbolometers configurations.

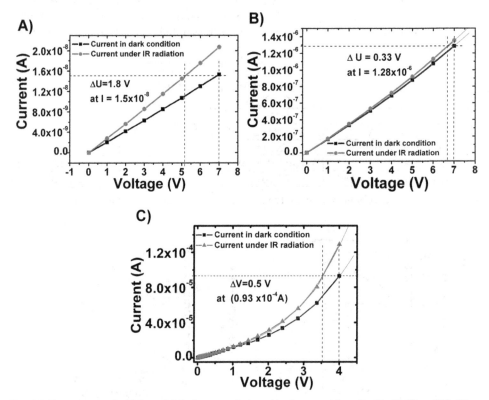

Fig. 11. Extraction of ΔU from I(U) characteristics: A) planar with a-Ge$_x$Si$_y$:H (Ge$_x$=0.5). B) planar with a-Ge$_x$Si$_y$B$_Z$:H (Ge$_x$=0.45, B$_z$=0.09). C) sandwich with a-Ge$_x$Si$_y$:H (Ge$_x$=0.5).

6.3 Noise spectral density measurements and detectivity calculations

Noise measurements in the microbolometers were performed with a lock-in amplifier ("Stanford Research Systems" - SR530). The noise of the system and the total noise (system + cell noise) were measured separately, and a subtraction of the system noise allowed us to obtain the noise of the device. The detectivity was calculated from the responsivity values and noise measurements. The current noise spectral density (NSD), I$_{cell\ noise}$ (f), of the fabricated devices with the different thermo-sensing films are shown in Fig. 12.

The NSD in the cell is obtained as $(I_{cell\ noise}(f))^2 = (I_{system\ +\ cell\ noise}(f))^2 - (I_{system\ noise}(f))^2$, where I$_{cell\ +\ system\ noise}$(f) is the NSD measured at the microbolometer with the measuring system and the I$_{system\ noise}$(f) is the NSD measured in the system without the microbolometer. In noise curves we observed different slopes at different frequencies and different cone frequencies. That data are shown in Table 6, where fc1 is the cone frequency 1, fc2 is the cone frequency 2, β is the slope of the curve in region 1 and γ is the slope of the curve in region 2.

The planar structure with a-Ge$_x$Si$_y$:H (Ge$_x$=0.5) film shows I$_{cell\ noise}$ (f) ≈ 10^{-16} AHz$^{-1/2}$; the planar structure with a-Ge$_x$Si$_y$B$_z$:H (Ge$_x$=0.45, B$_z$=0.09) film shows I$_{cell\ noise}$ (f) ≈ 10^{-14} AHz$^{-1/2}$; and the sandwich structure with a-Ge$_x$Si$_y$:H (Ge$_x$=0.5) has I$_{cell\ noise}$ (f) ≈ 10^{-11} AHz$^{-1/2}$.

The procedure for the detectivity calculation is shown in Eq. 19, where R$_I$ is the current responsivity, A$_{cell}$ is the detector area, I$_{noise}$ is the cell NSD and Δf = 1 is the bandwidth of the measurement system.

$$D^* = R_I\left(A_{cell}\right)^{1/2} / I_{noise} / (\Delta f)^{1/2} \tag{19}$$

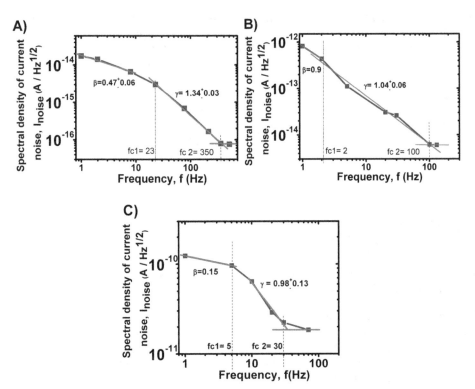

Fig. 12. Spectral density of current noise (NSD) of 3 microbolometers: A) planar with a-Ge$_x$Si$_y$:H (Ge$_x$=0.5). B) planar with a-Ge$_x$Si$_y$B$_z$:H (Ge$_x$=0.45, B$_z$=0.09). C) sandwich with a-Ge$_x$Si$_y$:H (Ge$_x$=0.5).

We calculated the detectivity values D* in the 3 structures. For the planar structure with the a-Ge$_x$Si$_y$:H (Ge$_x$=0.5) film we obtained D*= 7x10^9 cmHz$^{1/2}$W^{-1}; for the planar structure with the a-Ge$_x$Si$_y$B$_z$:H (Ge$_x$=0.45, B$_z$=0.09) film it is D*= 5.9x10^9 cmHz$^{1/2}$W^{-1}; and for the sandwich structure microbolometer with the a-Ge$_x$Si$_y$:H (Ge$_x$=0.5) film it is D*= 4x10^9 cmHz$^{1/2}$W^{-1}.

Samples	Frequency Regions				
	Region no. 1		Region no. 2		Region no. 3
	β	fc1 (Hz.)	γ	fc2 (Hz.)	Noise level (AHz$^{-1/2}$)
Planar structure a-Ge$_x$Si$_y$:H (Ge$_x$=0.5)	0.47	23	1.34	350	10^{-16}
Planar structure a-Ge$_x$Si$_y$B$_z$:H (Ge$_x$=0.45, B$_z$=0.09)	0.9	2	1.04	100	10^{-14}
Sandwich structure a-Ge$_x$Si$_y$:H (Ge$_x$=0.5)	0.15	5	0.98	30	10^{-11}

Table 6. NSD at different frequency regions in the different microbolometers structures.

7. Microbolometers thermal characterization and calibration curve

In order to estimate the temperature dependence of the thermal resistance of the microbolometers, I(U) measurements were performed in the range from 260 K to 360 K, as is shown in Fig. 13 A), where the bias is plotted as a function of current.

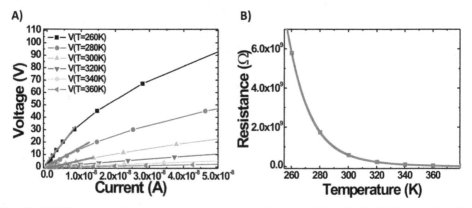

Fig. 13. A) U(I) curves of a planar structure microbolometer with a-Ge$_x$Si$_y$:H (Ge$_x$=0.5). B) Calibration curve of a planar structure microbolometer with a-Ge$_x$Si$_y$:H (Ge$_x$=0.5).

The slope of the linear part of each curve showed in Fig. 13 A) corresponds to the electrical resistance of the microbolometer for each temperature value (in the range of 260 K - 360 K). Once that is obtained the value of the electrical resistance for each value of temperature, it is possible to graph the electrical resistance of the microbolometer as a function of the temperature, also called the calibration curve, as is shown in Fig. 13 B).

The calibration curve is very important because from this curve is possible to calculate an increment in temperature in the microbolometer by measuring a change in its resistance.

The voltage-current curves in Fig. 13 A) have different resistance values as the current increases. The value of the resistance is affected by the temperature. Thus the resistance is calculated for each point of the Voltage-Current curves (obtained at different temperatures, 260, 270, etc.). If the resistance values obtained are compared with the calibration curve, it is possible to extract the increment of temperature (ΔT) for each point.

Fig. 14 A) shows the increment of temperature (ΔT) as a function of the power (P=U*I) applied to the microbolometer, for each temperature.

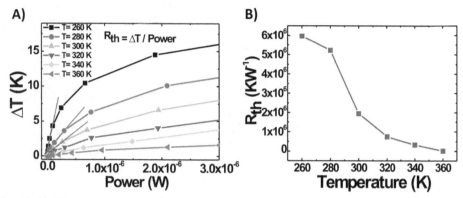

Fig. 14. A) ΔT vs Power curve of a microbolometer. B) Thermal resistance of a planar structure microbolometer with a-Ge$_x$Si$_y$:H (Ge$_x$=0.5).

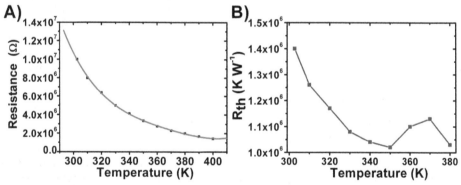

Fig. 15. Microbolometer with a-Ge$_x$Si$_y$B$_z$:H (Ge$_x$=0.45, B$_z$=0.09). A) Calibration curve. B) Thermal resistance.

The thermal resistance of the microbolometer, R$_{th}$, is then obtained as the slope of the increment of temperature in the microbolometer (ΔT) as a function of the power applied to it, for each temperature value. Fig. 14 B) shows the temperature dependence of the thermal resistance (R$_{th}$). Fig. 15 A) shows the calibration curve and Fig. 15 B) shows the temperature dependence of the thermal resistance (R$_{th}$) of the planar structure microbolometer with a-Ge$_x$Si$_y$B$_z$:H (Ge$_x$=0.45, B$_z$=0.09).

8. a-Ge$_x$Si$_y$:H and a-Ge$_x$Si$_y$B$_z$:H microbolometers compared with literature

The results obtained from the study of fabrication and characterization of different microbolometer structures, containing intrinsic a-Ge$_x$Si$_y$:H films and boron alloys a-Ge$_x$Si$_y$B$_z$:H, are discussed in the present section and compared with data reported in literature.

8.1 Thermo-sensing film characterization

Table 7 shows the most employed materials as thermo-sensing films in microbolometers. VO_x is one of the most employed materials (B. E. Cole, 1998), however this material is not compatible with Si CMOS standard technology and its TCR is not very large, around 0.021 K^{-1}. Amorphous $Ge_xSi_{1-x}O_y$ films have been employed in microbolometers (E. Iborra, 2002), these films are compatible with the CMOS technology and present a high TCR, around 0.042 K^{-1}; however also have a high resistance.

Material	TCR (K⁻¹)	Ea (eV)	σ_{RT} (Ω cm)⁻¹	Reference
VO_x	0.021	0.16	2×10^{-1}	B. E. Cole, 1998
a-Si:H (PECVD)	0.1 - 0.13	0.08-1	$\sim 1 \times 10^{-9}$	A. J. Syllaios, 2000
a-Si:H,B (PECVD)	0.028	0.22	5×10^{-3}	A. J. Syllaios, 2000
a-Ge_xSi_y:H (PECVD)*	0.043	0.34	1.6×10^{-6}	M. Moreno, 2008
$Ge_xSi_{1-x}O_y$	0.042	0.32	2.6×10^{-2}	E. Iborra, 2002

Table 7. Most common thermo-sensing materials employed in microbolometers.

At the present time a-Si:H and boron doped a-Si:H are employed in large microbolometer arrays (A. J. Syllaios, 2000). Intrinsic a-Si:H is compatible with CMOS technology and has a very high TCR, around 0.1-0.13 K^{-1}; however it is a highly resistive material, resulting in high resistive microbolometers which present a mismatch impedance with the readout circuits. Boron doped a-Si:H, has moderated resistivity, but also a reduced TCR, of 0.028 K^{-1}. Therefore none of those materials can be considered the optimum one as thermo-sensing material in microbolometers. Intrinsic a-Ge_xSi_y:H films presents a large TCR, around 0.043 K^{-1}, a moderated resistivivy and is compatible with the Si CMOS technology; those characteristics make this material suitable as thermo-sensing film for microbolometer arrays, however the resistivity is still an issue.

Amorphous germanium-silicon-boron alloys a-$Ge_xSi_yB_z$:H, have been studied in order to reduce the high resistivity presented in intrinsic films. From the conductivity characterization in the thermo-sensing films, we can state that the a-$Ge_xSi_yB_z$:H alloys demonstrated an increment in their conductivity (between 2 and 3 orders of magnitude) in comparison with that of the intrinsic a-Ge_xSi_y:H film. However the increment in σ_{RT} was accompanied by a reduction in TCR, to above 0.028 K^{-1}.

The deposition rate in the boron alloys is above 2-3 times larger than that of the intrinsic film. Thus B incorporation during the thermo-sensing deposition, enhance the deposition rate. The deposition of the thermo-sensing films over a SiN_x micro-bridge, has as consequence a reduction in the film conductivity, the stress arisen in the SiN_x micro-bridge could be the cause for the σ_{RT} reduction. The a-$Ge_xSi_yB_z$:H films compared with the another thermo-sensing materials, have better performance characteristics, which are: compatibility with the Si CMOS technology, moderated values of TCR, comparables with those of the VO_x and a-Si:H films, and reduced resistivty. In general the a-$Ge_xSi_yB_z$:H alloys have similar characteristics than those of the the a-Si:H,B thermo-sensing film (A. J. Syllaios, 2000), but also have one order of magnitude shorter resistivity.

8.2 Microbolometers characterization

Table 8 shows the main performance characteristics of the microbolometers reported in literature, which are compared with the different microbolometer configurations studied in this work, containing intrinsic and boron alloys thermo-sensing films. Commercial planar structure microbolometers based on VO$_x$ films (B. E. Cole, 1998), present moderated values of TCR, around α=0.021 K^{-1} and high values of voltage responsivity, R$_U$=2.5x10^7 VW^{-1}. Another performance characteristics as detectivity, D* are not published. The main drawback of these devices is their un-compatibility with Si CMOS technology, thus special installations are necessary for their fabrication, which make impossible to fabricate those devices in any standard Si CMOS fabrication line.

Thermo sensing film	E$_a$, eV	TCR,α K^{-1}	Cell area, A$_{cell}$ μm²	Cell resistance, R$_{cell}$, Ohm	Voltage responsivity R$_U$, VW^{-1}	Current responsivity, R$_I$, AW^{-1}	Detectivity, D* cmHz$^{1/2}$W^{-1}	References
VO$_X$	0.16	0.021	50 x 50	-	2.5 x 10^7	-	-	B. E. Cole, 1998
Ge$_x$Si$_{1-x}$O$_y$	0.32	0.042	50 x 50	7x10^5	1x10^5	-	6.7x10^8	E. Iborra, 2002
a-Si:H,B	0.22	0.028	48 x 48	3 x10^7	10^6	-	-	A. J. Syllaios, 2000
a-Si:H	0.8-1	0.039	25 x 25	>10^9	-	-	-	A. J. Syllaios, 2000
a-Ge$_x$Si$_y$:H Ge$_x$=0.5	0.34	0.043	70 x 66	5x10^8	7.2x10^5 *	2x10^{-3}	7.9x10^9	Planar M. Moreno, 2008
a-Ge$_x$Si$_y$B$_z$:H Ge$_x$=0.45, B$_z$=0.09	0.21	0.027	70 x 66	1x10^6	1.2x10^5 *	3x10^{-2}	5.9x10^9	Planar M. Moreno, 2008
a-Ge$_x$Si$_y$B$_z$:H Ge$_x$=0.55, B$_z$=0.07	0.20	0.028	70 x 66	3x10^6	1.8x10^5 *	7x10^{-2}	2x10^9	Planar M. Moreno, 2008
a-Ge$_x$Si$_y$:H Ge$_x$=0.5	0.34	0.043	70 x 66	1x10^5	2.2x10^5 *	0.3 - 14	4x10^9	Sandwich M. Moreno, 2008

Table 8. Comparison of characteristics of micro-bolometers with literature.

Planar structure microbolometers based on resistive a-Si:H, present high values of TCR, around α=0.1-0.13 K^{-1}, however also have very high values of resistance. Comercial a-Si:H,B based planar structure microbolometers (A. J. Syllaios, 2000) have moderated values of TCR, around α=0.28 K^{-1}, a cell resistance R$_{cell}$ = 3x10^7 Ω, and a high voltage responsivity, around R$_U$= 10^6 VW^{-1}. However values of D* are not reported.

Microbolometers based on a-Ge$_x$Si$_{1-x}$O$_y$ (E. Iborra, 2002) have demonstrated high values of TCR, around α=0.042 K^{-1}, a moderated cell resistance, R$_{cell}$ = 7x10^5 Ω, and moderated detectivity D*=6.7 x10^8 cmHz$^{1/2}$W^{-1}.

In our work, the planar structure microbolometers based on intrinsic a-Ge$_x$Si$_y$:H films have a very high value of TCR, around α=0.043 K^{-1}, a current responsivity, R$_I$=2x10^{-3} AW^{-1}, a very low current NSD, $I_{noise} \approx 1 \times 10^{-15}$ AHz$^{-1/2}$, resulting in a very high detectivity D*= 7.9×10^9 cmHz$^{1/2}$W^{-1}. However those devices still have a high cell resistance, $R_{cell} = 5 \times 10^8$ Ω.

The boron alloy planar structure microbolometers have a cell resistance, around $R_{cell} \approx (1-3)$ $\times 10^6$ Ω, which is two orders of magnitude shorter than that of the planar structure devices with intrinsic film and one order of magnitude shorter than that of the a-Si:H,B commercial devices (A. J. Syllaios, 2000). The current responsivity is around $R_I= (3-7) \times 10^{-2}$ AW^{-1}, and the current NSD, $I_{noise} \approx 10^{-13}$ AHz$^{-1/2}$, which results in a high detectivity D*= (2-6) $\times 10^9$ cmHz$^{1/2}$W^{-1}.

The sandwich structure microbolometer with the intrinsic a-Ge$_x$Si$_y$:H film, presents the shortest cell resistance of the devices reported in literature, $R_{cell} \approx 1 \times 10^5$ Ω, which is 3 orders of magnitude less than that of the planar devices with the same intrinsic film; one order of magnitude shorter than that of the boron alloy devices; 2 orders shorter than that of the a-Si:H,B devices; and near to 1 order of magnitude shorter than that of the a-Ge$_x$Si$_{1-x}$O$_y$ microbolometers. The TCR in sandwich structures is very high, around α=0.043 K^{-1}, the current responsivity is in the range of R_I= (0.3 -14) AW^{-1}, which is around 2 - 3 orders of magnitude larger than that of the boron alloys (a-Ge$_x$Si$_y$B$_z$:H) planar structure microbolometers and around 3 - 4 orders of magnitude larger than the intrinsic a-Ge$_x$Si$_y$:H film planar structure devices. However the sandwich structure presents a larger current NSD, $I_{noise} \approx 10^{-11}$ AHz$^{-1/2}$, which results in a detectivity D*= 4 $\times 10^9$ cmHz$^{1/2}$W^{-1}.

9. Conclusion

Uncooled microbolometers are reaching performance levels which previously only were possible with cooled infrared photon detectors. For uncooled infrared bolometer arrays based on amorphous silicon films the efforts have been conducted to increase the number of pixels included in the arrays, rather than improve the performance characteristics of the microbolometers. Plasma deposited amorphous germanium-silicon (a-Ge$_x$Si$_y$:H) and amorphous germanium-silicon-boron (a-Ge$_x$Si$_y$B$_z$:H) used as thermo-sensing films provided a high TCR and, as a consequence, a high responsivity and high detectivity with a improved conductivity. Thus a-Ge$_x$Si$_y$:H and a-Ge$_x$Si$_y$B$_z$:H are very promising materials for its integration on IR detector arrays, and its circuitry in the same chip, avoiding the problems of matching with the input impedance of the electronic circuits. Moreover the manufacture of those devices is aligned with standard CMOS and MEMS foundry processes.

10. Acknowledgment

The authors acknowledge: 1. CONACYT for the support for this research through the grant of projects no. D48454-F and 154112. 2. Dr Y. Kudriavtsev from CINVESTAV, Mexico, for SIMS characterization. 3. INAOE, Mexico, for the permission for reproduction of some figures from the Ph.D. thesis work of M. Moreno, titled "Study of IR un-cooled micro-bolometers arrays based on thin films deposited by plasma".

11. References

Ambrosio, R.; Torres, A.; Kosarev, A.; Illinski, A.; Zuniga, C.; Abramov, A. (2004). Low frequency plasma deposition and characterization of Si$_{1-x}$Ge$_x$:H:F films, *Journal of Non-crystalline Solids*, Vol. 338-340, pp. 91-96, ISSN 0022-3093.

Cole, B. E.; Higashi, R. E.; Wood, R. A. (1998). Monolithic Two-Dimensional Arrays of Micromachined Microstructures for Infrared Applications, *Proceedings of the IEEE*, Vol. 86, No. 8, pp. 1679 -1686. ISBN 0018-9219, August , 1998.

Cole B. E.; Higashi R.E; Wood, R. A. (2000). Micromachined Pixel Arrays Integrated with CMOS for Infrared Applications, *Proceedings of IEEE International Conference on Optical MEMS 2000*, pp. 63 – 64, ISBN 0-7803-6257-8, Kauai, HI , USA, August 2000.

Delerue, J.; Gaugue, A.; Testé, P.; Caristan, E.; Klisnick, G.; Redon, M.; Kreisler, A. (2003). YBCO Mid-Infrared Bolometer Arrays, *IEEE Transactions on Applied Superconductivity*, Vol. 13, No. 2, pp. 176-179. ISSN 1051-8223.

Hirota, M.; Morita, S. (1998). Infrared sensors with precisely patterned Au-black absorption layer, *Proceedings of SPIE Infrared Technology and Applications XXIV*, Vol. 3436, pp. 623-634. ISBN 9780819428912, San Diego Ca. USA, July 19-24, 1998.

Iborra, E.; Clement, M.; Vergara Herrero. L.; Sangrador, J. (2002). IR uncooled bolometers based on amorphous Ge$_x$Si$_{1-x}$O$_y$ on Silicon Micromachined structures, *Journal of Microelectromechanical Systems*, Vol. 11, No. 4, pp. 322- 329. ISSN 1057-7157.

Kosarev, A.; Moreno, M.; Torres A., Ambrosio R. (2006). Un-cooled micro-bolometer with Sandwiched Thermo-sensing Layer Based on Ge films deposited by Plasma, *Proceedings of Materials Research Society - Amorphous and Polycrystalline Thin-Film Silicon Science and Technology-2006*. Vol. 910, A17-05, ISBN 978-55899-866-7, Warrendale, PA, USA, April 2006.

Kruse, P. W. (2001). Uncooled thermal Imaging, arrays systems and applications, *Tutorial text in optical engineering*, Volume TT51, SPIE Press, ISBN 9780819441225, Bellingham, Washington USA.

Moreno, M.; Kosarev, A.; Torres, A.; Ambrosio, R. (2007). Fabrication and Performance Comparison of Planar and Sandwich Structures of Micro-bolometers with Ge Thermo-sensing layer. *Thin solid films*, Vol. 515, pp. 7607-7610, ISSN 0040-6090.

Moreno, M.; Kosarev, A.; Torres, A.; Ambrosio, R. (2008). Comparison of Three Un-Cooled Micro-Bolometers Configurations Based on Amorphous Silicon-Germanium Thin Films Deposited by Plasma, *Journal of Non Crystalline Solids*, Vol. 354, pp. 2598-2602, ISSN 0022-3093.

Moreno, M.; Kosarev, A.; Torres, A.; Ambrosio, R.; Garcia, M.; Mireles, J. (2010). Measurements of thermal characteristics in silicon germanium un-cooled micro-bolometers. *Physica Status Solidi C*, C 7, No. 3–4, pp. 1172– 1175, ISSN 1610-1634.

Pitigala, P.K.D.D.P.; Jayaweera, P.V.V.; Matsik, S.G.; Perera, A.G.U.; Liu H.C. (2011). Highly sensitive GaAs/ AlGaAs heterojunction bolometer, *Sensors and Actuators A*, Vol. 167 pp. 245–248, ISSN 0924-4247.

Rogalski, A. (2003). Infrared detectors: status and trends, *Progress in Quantum Electronics*, Vol. 27, pp. 59-210. ISSN 0079-6727.

Syllaios, A. J.; Schimert T. R.; Gooch, R. W.; Mc.Cardel, W. L.; Ritchey, B. A.; Tregilgas, J. H. (2000). Amorphous silicon microbolometer technology, *Proceedings of Maerials Research Society - Amorphous and Heterogeneous Silicon Thin Films 2000*, Vol. 609, A14.4, ISBN 9781558995178, San Fco. Cal. USA, April, 2000.

Schaufelbühl, A.; Schneeberger, N.; Münch, U.; Waelti, M.; Paul, O.; Brand, O.; Baltes, H.; Menolfi, C. (2001). Uncooled Low-Cost Thermal Imager Based on Micromachined

CMOS Integrated Sensor Array, *Journal of Microelectromechanical systems*, Vol. 10, No. 4, pp. 503-510. ISSN 1057-7157.

Sedky, S.; Fiorini, P.; Caymax, M.; Baert, C.; Hermans L.; Mertens, R. (1998). Characterization of Bolometers Based on Polycrystalline Silicon Germanium Alloys, *IEEE Electron Device Letters*, Vol. 19, No. 10, pp. 376- 378. ISSN 0741-3106.

Tanaka, A.; Matsumoto, S.; Tsukamoto, N.; Itoh, S.; Chiba, K.; Endoh, T.; Nakazato, A.; Okuyama, K.; Kumazawa, Y.; Hijikawa, M.; Gotoh, H.; Tanaka, T.; Teranishi, N. (1996). Infrared focal plane array incorporating silicon IC process compatible bolometer, *IEEE Transactions on Electron Devices*, Vol. 43, Issue 11, pp. 1844 – 1850, ISSN 0018-9383.

Takami, H.; Kawatani, K.; Kanki, T.; Tanaka, H. (2011). High Temperature-Coefficient of Resistance at Room Temperature in W-Doped VO_2 Thin Films on Al2O3 Substrate and Their Thickness Dependence, *Japanese Journal of Applied Physics*, Vol. 50, pp. 055804-1 - 055804-3, ISSN 0021-4922.

Torres, A.; Moreno, M.; Kosarev, A.; Heredia, A. (2008). Thermo-sensing Germanium-Boron-Silicon Films Prepared by Plasma for Un-cooled Micro-bolometers, *Journal of Non Crystalline Solids*, Vol. 354, pp. 2556-2560, ISSN 0022-3093.

2

Group IV Materials for Low Cost and High Performance Bolometers

Henry H. Radamson[1] and M. Kolahdouz[2]
[1]School of Information and Communication Technology, KTH
Royal Institute of Technology, Kista
[2]Thin Film Laboratory, Electrical and Computer Engineering Department,
University of Tehran, Tehran,
[1]Sweden
[2]Iran

1. Introduction

Infrared (IR) imaging has absorbed a large attention during the last two decades due to its application in both civil and military applications (Per Ericsson et al., 2010; Lapadatu et al., 2010; Sood et al., 2010). Thermal detector is presently revolutionizing the IR technology field and it is expected to expand the market for cameras. These detectors are micro-bolometers and are manufactured through micro-maching of a thermistor material. Since these detectors demand no cryogenic cooling, they provide the opportunity for producing compact, light-weight, and potentially low-cost cameras. The preferred functioning wavelength regions for these detectors are usually 8–12 μm due to the high transparency of the atmosphere in these regions.

Micro-bolometers function through absorption of infrared radiation on a cap layer which warms the bolometer's body and raises the temperature. This temperature change is sensed by a thermistor material integrated in the bolometer, i.e. a a material whose resistivity changes with temperature variation. The whole detector body consists of a thin membrane which is thermally isolated and is fastened to the wafer via two thin legs. The legs are connected to a CMOS-based read-out integrated circuit (ROIC). A thin oxide or nitride layer is deposited to ensure the stability of the legs in contact to the ROIC body (see Fig. 1). The whole detector is vacuum encapsulated to reduce effectively the thermal conductance. Signal processing is obtained and multiplexing electronics (CMOS) is integrated within the silicon substrate. All the membranes are in form of pixels which are bonded to a read-out circuit to amplify the generated signal (Kvisterøy et al., 2007; J. Källhammer et al., 2006; F. Niklaus, Kälvesten, & G. Stemme, 2001; F. Niklaus, Vieider, & Jakobsen, 2007).

This chapter will present the benefits and drawbacks of group IV thermistor materials in bolometers. The proposed structures are composed of multi-quantum wells (MQWs) or dots (MQDs), structures of Si(C) (barrier)/SiGe(C) (quantum well layer) and their combination with a Schottky diode.

1.1 Thermal detectors

A detector may be simply represented by a thermal capacitance C_{th} coupled via the thermal conductance G_{th} to a heat sink at the constant temperature T. When the detector is exposed to radiation, the temperature variation can be calculated through the heat balance equation. For any thermistor material assuming periodic radiant power, temperature variation is given by (Kruse, McGlauchlin, & McQuistan, 1962; Smith, Jones, & Chasmar, 1968):

$$\Delta T = \frac{\varepsilon \Phi_0}{\left(G_{th}^2 + \omega^2 C_{th}^2\right)^{\frac{1}{2}}} \tag{1}$$

where ΔT is the optically induced temperature variation due to the incident radiation Φ ($\Phi_0 \exp(i\omega t)$) and ε is the emissivity of the detector. The usual procedure employed in bolometer detectors to achieve a good IR absorption is depositing a transparent thin metallic film on top of the device. Free electron absorption in metal films guarantees the absorption of about 50% of the incident IR radiation (Liddiard, 1984). In order to further enhance infrared absorbance, a resonant cavity is employed in the detector structure. The resonant cavity involves an absorbing membrane suspended at a distance d above the cavity reflector metal. The resonant absorbance peaks correspond to the condition for minimum reflectance. The three resonance absorbance peaks are $\lambda/4$, $3\lambda/4$ and $5\lambda/4$ in the LWIR spectral band (Schimert et al., 2008).

A more practical design for a bolometer with high performance is to create the cavity within the sensor membrane itself. In this case, a reflective area is (a mirror-like) deposited on the bottom side of the bolometer membrane (Per Ericsson et al., 2010).

The LWIR radiation is within 8-12 µm wavelength region and the maximum absorption is obtained when the total bolometer membrane thickness of suspended membrane including the absorbant cap layer, thermistor material and the reflector layer is ~2-3 µm. This thickness is a rough estimation since the semiconductor thermistor material consists of a multilayer (e.g. Si/SiGe) structure. Thus, the final optimized membrane thickness has to be obtained by optical simulations considering the optical properties of all layers.

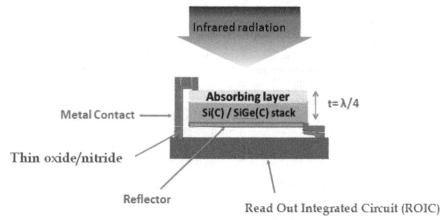

Fig. 1. A schematic cross-section of a bolometer pixel for an optimal absorption.

It is worth mentioning here that for a bolometer, the membrane thickness will affect not only the optical properties, but also the bolometer mass and the electrical resistance. Therefore, it is important to take into account both optical and electrical factors when performing the calculations.

In order to increase the temperature response of the detector, the thermal capacity of the detector (Cth) and the thermal coupling to its surroundings (Gth) must be as small as possible. The thermal contacts of the detector with surroundings should be reduced while the interaction with the incident beam must be optimized. In practice, the detectors are vacuum encapsulated to become thermally isolated. The thermal response time (τth) for such a detector can be written as:

$$\tau_{th} = \frac{C_{th}}{G_{th}} = C_{th} R_{th} \qquad (2)$$

The typical response time for a thermal detector is in millisecond range which is longer than that of photon detectors (microsecond range). Eq.1 can then be rewritten as:

$$\Delta T = \frac{\varepsilon \Phi_0 R_{th}}{\left(1 + \omega^2 \tau_{th}^2\right)^{\frac{1}{2}}} \qquad (3)$$

This means that the detector sensitivity is higher for lower frequency range. The voltage responsivity of the detector is given by the ratio of the output voltage signal (V_s) to the input radiation power (Φ_0):

$$R_V = \frac{V_s}{\Phi_0} = \frac{K \Delta T}{\Phi_0} \qquad (4)$$

where the generated output voltage is assumed to be linearly proportional to the temperature difference and K is linearly dependent on the thermistor TCR value. Substituting eq.3 in eq.4 results in the following equation (Liddiard, 1984):

$$R_V = \frac{V_s}{\Phi_0} = \frac{|\alpha| I_s R_b R_L \varepsilon R_{th}}{(R_b + R_L)^2 \left(1 + \omega^2 \tau_{th}^2\right)^{\frac{1}{2}}} \qquad (5)$$

It can be deduced from the final expression that at low frequencies ($\omega \ll 1/\tau$), the responsivity is proportional to the thermal resistance of the detector (R_{th}) and not the thermal capacitance. This is exactly the opposite at high frequencies. As the operating frequency increases beyond the cut-off frequency ($f = 1/2\pi\tau$) the responsivity of the detector rapidly declines. Thus, good responsivity can be achieved by using a high TCR thermistor which is a characteristic of semiconductors rather than metals, and by minimizing G_{th} through a good thermal isolation of the bolometer.

Thus designing a high quality IR camera is not an easy task and many other parameters and issues e.g. thermistor material choice, costs and feasibility have to be well thought of. Meanwhile a more physical discussion for bolometers should cover the wavelength dependence of the interaction between the optical absorption and the bolometer mass, and the black body radiance. A more interpretable image of a resistive bolometer can be expressed by the noise equivalent temperature difference (NETD) as follows (Frank Niklaus, Decharat, Jansson, & Göran Stemme, 2008):

$$NETD = \frac{4F^2CV_N\left(1+\omega^2\tau_{th}^2\right)^{1/2}}{\tau_{th}\,U_{bias}\,TCR\,\pi\,A_{bolo}\dfrac{\delta}{\delta T_{obj}}\int_{\lambda_1}^{\lambda_2}\phi(\lambda)\varepsilon(\lambda)L\left(\lambda,T_{obj}\right)d\lambda}[K]\qquad(6)$$

where λ is the wavelength and T_{obj} the object temperature, L the black body radiance, ε the bolometer absorption, φ the wavelength dependent transmission of the optical system, TCR the temperature coefficient of resistance, U_{bias} bias voltage applied to the thermistor, τ the thermal time constant of the membrane, ω the image modulation frequency, V_N the RMS noise voltage, F is the f-number of the optical system, and C the heat capacity of the membrane.

This expression indicates that the thickness cannot be freely adjusted to obtain the optical $\lambda/4$ cavity and a larger C increases the NETD.

1.2 Figures of merit for thermistor materials

The figures of merit for a thermistor are temperature coefficient of resistance (TCR) and signal noise level. Today, commonly used thermistor materials such as vanadium oxide (VO_x), amorphous, and polycrystalline semiconductors demonstrate moderate noise levels and TCR values around 2%–4% (Lv, Hu, Wu, & Liu, 2007; Moreno, Kosarev, Torres, & Ambrosio, 2007). Recent studies have proposed single crystalline (sc) SiGe as a thermistor material (Di Benedetto, Kolahdouz, Malm, Ostling, & H. H. Radamson, 2009; Vieider et al., 2007; S. Wissmar, H. Radamson, Kolahdouz, & J. Y. Andersson, 2008) demonstrating a high signal-to-noise level. This has been achieved by high epitaxial quality and smooth interfaces between the Si and SiGe layers. The simulations of the fully strained SiGe/Si quantum well structure indicate that the TCR performance can be improved to 6%–8% for 70%–100% Ge in sc-SiGe layers. Although these predicted values for sc-SiGeseem to be outstanding, but so far no experimental data have confirmed the theoretical calculations. One obstacle to overcome is the strain relaxation of the epitaxial SiGe layers which results in surface roughness (Di Benedetto et al., 2009). Since the properties of this thermistor material is improved by increasing the Ge content, producing a high quality SiGe with high Ge content (>35%) is a challenging effort.

1.3 Temperature response and sensitivity

The intrinsic part of a bolometer consists of the "thermistor" material. This part responds to temperature variations which result in resistivity changes. The criteria for an ideal thermistor material can be addressed as follows (Schimert et al., 2008): 1) a high temperature coefficient of resistance (TCR); 2) a high signal–to–noise ratio (SNR); 3) a sufficiently low thermal response time constant which leads to a high responsivity; 4) the ability to form a thermally isolated optical cavity from the material; 5) the mature material growth technology that is compatible with integration on a substrate containing the VLSI signal processing functions; and 6) the possibility to manipulate a wide range of bolometer resistance.

Temperature Coefficient of Resistance for a thermistor material is the parameter used to quantify the temperature sensitivity and it is defined as (Di Benedetto et al., 2009):

$$\alpha = \frac{1}{R}\frac{\partial R(T)}{\partial T}[K^{-1}] \tag{7}$$

The resistivity is the exponential function of thermal activation conductance which is expressed by:

$$\rho = \rho_0 \exp(\frac{E_a}{kT}) \tag{8}$$

where ρ, ρ_0, E_a and k are the resistivity, the measured pre-factor, the activation energy and Boltzmann's constant. In semiconductors, α can be expressed by the activation energy derived from Arrhenius plot (Schimert et al., 2008):

$$\alpha = -\frac{1}{kT^2}[\frac{3}{2}k_B T + E_f - V] \tag{9}$$

The thermistor materials have either positive temperature coefficient of resistance (PTC) or negative temperature coefficient of resistance (NTC). The first group includes materials like metals in which the resistance increases with increasing the temperature; whereas, the latter group are composed of semiconductor materials in which the resistance decreases with increasing the temperature.

For a Si(C)/SiGe(C) MQW structure, E_a becomes the barrier height V (see Fig. 2). In order to maximize the TCR of bolometers, high Ge content (or even pure Ge) on Si is required.

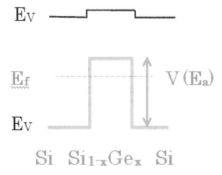

Fig. 2. A schematic drawing of the banddiagram of Si/SiGe/Si heterojunctions (type II-alignment).

However, due to lattice mismatch of ~4% between Si and Ge strain relaxation occurs when the thickness of the SiGe layers exceeds a critical value (Bean, Feldman, Fiory, Nakahara, & Robinson, 1984). High quality SiGe quantum wells are grown when the layer thickness is a value within the meta-stable region (see Fig. 3a). Otherwise the strain relaxation for the thin SiGe layers is monitored through interfacial roughness and no dislocations are observed (see Fig. 3b).

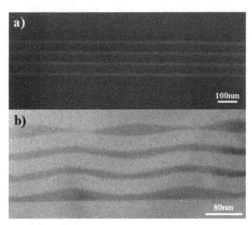

Fig. 3. The cross-sectional HRSEM view of a) $Si_{0.72}Ge_{0.28}$ / Si and b) $Si_{0.68}Ge_{0.32}$ / Si stack grown at 650 °C. The dark strips are SiGe layers.

For a thin SiGe well, nevertheless, the ground state in the well will shift away from the valence band edge because of carrier confinement (Cohen-Tannoudji, Diu, & Laloe, 1977). In order to calculate the energy levels and sub-bands in terms of quantum well's profile (Ge content and layer thickness) the Schrödinger equation has to be solved for holes in the valence band using 6*6 Luttinger Hamiltonian. These theoretical calculations can be essential for designing structures for high TCR values.

1.4 Signal-to noise ratio

Noise for an electronic signal is a stochastic random variation which makes it difficult to distinguish the signal amplitude and as a result the IR detectivity becomes limited. The noise occurs when the voltage or current measurements are performed for a device and it is the summation of contributions from the different sources. These sources can be divided into two main groups; a) external or extrinsic sources which are generated due to the surrounding performance of the device and b) Internal or intrinsic sources which refers to random fluctuation in the carrier transport because of defects and imperfections in the device structure.

The thermal noise (it is also so-called Nyquist or Johnson noise) is similar as Brownian motion of the charged carriers in a material and its nature is a random thermal motion. In a conductive material, at non-zero temperature T, electrons vibrate randomly depending on T. This noise is expressed for a resistor R in a Δf bandwidth by:

$$V_j^2 = 4kTR\Delta f \tag{10}$$

where k is the Boltzmann constant. This means that the thermal noise can be minimized for bolometer application by lower resistive material, lower operating temperature and narrower bandwidth. However, an actual bolometer application requires a finite limit to the bandwidth through the scanning and readout of the detectors and ambient temperature operation. Thermal or temperature fluctuation noise is another noise source which must be discussed to evaluate the detectivity of the device.

Since the detector consists of an array of pixels which are suspended membranes, and has legs connected to the ROIC, then a thermal or temperature fluctuation noise is created which affects the detectivity of the device. This will transform in a form of an electric noise because of the coupling between the temperature and the resistance. Temperature fluctuation is expressed as follows (Kruse et al., 1962; Smith et al., 1968):

$$V_{th}^2 = \frac{4kT^2\Delta f}{\left(1 + \omega^2\tau_{th}^2\right)^{1/2}}K^2R_{th} \tag{11}$$

The other important source of noise for IR detection is the "background noise". Heat exchange due to radiation between the detector at temperature T_d and the environment at temperature T_b generate voltage noise which is so called "background noise". As an example an exchange of the heat between the sensitive area of the detector and the surrounding substrate and contact legs (which are in thermal contact with the detector) introduces a random fluctuation in the temperature. This will then transform into a kind of electric noise because of the coupling between the temperature and the resistance. The expression is given by (Kruse et al., 1962; Smith et al., 1968):

$$V_b^2 = \frac{8k\varepsilon\sigma A(T_d^2 + T_b^2)}{1 + \omega^2\tau_{th}^2}K^2R_{th}^2 \tag{12}$$

where ε is emissivity, σ is Stefan-Boltzmann constant and A is the area of the detector. For semiconductor thermal detectors, $1/f$ noise or Flicker Noise is the most predominant noise at low frequency. $1/f$ noise can be evaluated by noise constant: $K1/f = \gamma/N$ where γ is known as the Hooge's constant and N is the total number of free charges (Hooge, 1994). The exact description for the origin of $1/f$ noise is not clear but the interactions of carriers with defects, surface states and other events (e.g. recombination and trapping-detrapping) are the major causes of this noise in semiconductors. In the case of bolometers, a simple expression for voltage power spectrum density (PSD) can be written as follows:

$$S_V = \frac{K_{1/f}V_{bias}^\beta}{f^\gamma} \tag{13}$$

where $K_{1/f}$ is a noise constant. In a similar way, $1/f$ noise voltage can be written as:

$$V_{1/f}^2 = \frac{K_{1/f}I^\beta}{f^\gamma}\Delta f \tag{14}$$

The parameters γ, β and $K_{1/f}$ in equations 13 and 14 are dependent on the material, processing, metal contacts and surfaces and thus, very difficult to calculate analytically (Hooge, 1994). Since $1/f$ noise relates to defects and imperfections in the active part of the bolometer, it is believed that using single-crystalline (sc) materials will demonstrate low noise constant in comparison to polycrystalline or amorphous ones. Thus, a solution for increasing the D* of a bolometer is to use mono-crystalline temperature sensing bolometer materials with a low $1/f$ noise constant (Kolahdouz, Afshar Farniya, Di Benedetto, & H. Radamson, 2010).

For most bolometer applications, the frequency exponent γ in equations 13 and 14 is close to 1. The square of total noise voltage for a thermistor material in active part of a bolometer may be formulated in eq as:

$$V_J^2 = V_J^2 + V_{th}^2 + V_b^2 + V_{1/f}^2 \qquad (15)$$

When a thermal detector absorbs the electromagnetic radiation, both output signal and noise will be generated. High amplitude output signal and low noise level are desired in an infrared detector. To evaluate the performance of the detector, "Detectivity" may be defined as follow:

$$D = \frac{R_V}{\Phi_0} = \frac{V_s}{\Phi_0 V_n} \qquad (16)$$

where V_n, V_s and Φ_n are RMS signal voltage, noise voltage, and incident power respectively.

The detectivity is proportional to detector area and electrical bandwidth. Therefore, the normalized detectivity D* is given by:

$$D^* = D \times A^{1/2} \times \Delta f^{1/2} \qquad (17)$$

In a thermal detector, D* can be expressed as:

$$D^* = \frac{K\varepsilon R_{th}A^{1/2}}{\left(1+\omega^2\tau_{th}^2\right)^{1/2}(V_J^2 + V_{th}^2 + V_b^2 + V_{1/f}^2)^{1/2}} \qquad (18)$$

where A is the pixel area. From eq. 16, it can be concluded that the detectivity may be enhanced by increasing the responsivity and/or decreasing the noise. The responsivity, like the Flicker noise, increases linearly with voltage, while the Johnson noise is independent of voltage. At small voltages, the noise is mainly Johnson noise. But, at sufficiently high V, noise is dominated by the Flicker noise and D* is independent of voltage. According to the previous calculations, the highest detectivity for a thermal detector at room temperature and viewing background at room temperature is about 2×10^{10} cmHz$^{1/2}$W^{-1} which can be referred to as the thermal detectors theoretical limitation. The published photon detectors have shown higher detectivities as a result of their limited spectral responses.

In addition to the above discussions, the importance of the electrical contacts' influence on the thermistor's performance has to be emphasized. The current-voltage characteristics of the thermistor materials are greatly influenced by the nature of the metal/silicide-semiconductor interface. Ohmic contact with low contact resistance is the requirement for low noise level for many applications. However, when large electrical current is involved a low sheet resistance contact is required to make the current flow uniform without localized overheating. A metal with low work function will form an ohmic contact to an n-type semiconductor with surface states. The reverse story is true for a p-type semiconductor. In these cases, introducing higher doping concentration reduces the contact resistance near the contact surface (barrier thinning).

2. Noise measurement of the thermistor materials

Power spectral density (PSD) of voltage noise is measured for different pixels. The measurements can be performed at different temperatures inside a shielded probe station to avoid light and the environmental noise. The frequency range is usually 0.3-10,000 [Hz] with some sub-intervals. Each final PSD vs frequency curve includes many thousands of data. Fig. 4 shows the experimental set up. The device is biased through a circuit isolated by a metallic box.

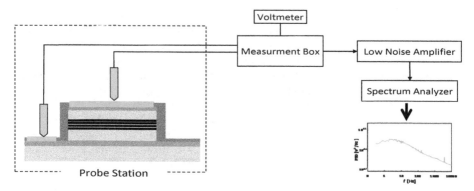

Fig. 4. Experimental set up for measuring PSD of voltage noise.

3. Thermistor materials for uncooled bolometers

Bolometers are uncooled detectors and today they have dominantly taken over the IR market. Among the existing thermistor materials, Vanadium oxide (VO_x) is mainly used for bolometer applications and its performance has been studied and improved during many years. Nowadays, it is believed that VO_x technology will be challenged in the near future by the new silicon based materials due to their low cost structure, and easier manufacturability (Rogalski, 2011).

In this part, an overview of the VO_x material properties is presented and later, the discussions will be extended towards single-crystalline Si-based materials.

3.1 Vanadium oxide

Vanadium oxides are the most popular thermistor material in fabrication of today's IR detectors. This material is grown by different techniques e.g. sputtering (Y. Han et al., 2003; Lv et al., 2007; Moon, Y. Han, K. Kim, S. Lee, & Shin, 2005), reactive e-beam evaporation (Subrahmanyam, Bharat Kumar Redd, & Nagendra, 2008), reactive e-beam evaporation (H. Wang, Yi, & Chen, 2006), PLD (Kumar et al., 2003), and CVD (Mathur, Ruegamer, & Grobelsek, 2007). Many reports demonstrate that by tuning the growth parameters, a transition occurs from amorphous to nano-crystalline FCC VOx ($0.8 < x < 1.3$) (Cabarcos et al., 2011).

Since vanadium atom has a half-filled d-shell, there exist a set of valence states to form a number of oxide phases. The typical phases are known as VO, V_2O_3, VO_2 (or V_2O_4) and V_2O_5 (or as a "mixed oxide) (Subrahmanyam et al., 2008).

Among these phases, V_2O_3 shows semiconductor-to-metal transition at ~160K and demonstrates a very low resistance. For bolometer application, VO_2 phase is typically used due to its high TCR value but its metal transition temperature occurs at ~341 K which restricts the bolometer's IR detection. Another interesting vanadium oxide is V_2O_5. This phase shows a good TCR; however, its resistance is very high resulting in high noise value.

Thus, a mixed phase of VO_2 and V_2O_5 may demonstrate an appropriate resistivity which is convenient and matches also with the readout electronics for high sensitive bolometers (Malyarov, 1999).

The growth of VO_x films requires extra care since the morphology of the film is sensitive to the growth parameters. In most cases, the substrate temperature and the oxygen pressure are the two crucial growth parameters to control the composition and the grain size of the oxide films. The grown VO_x films demonstrate TCR values in range of 2 -3%K^{-1}. Some of the published data are addressed in table 1.

Technique	Material	Processing temperature (K)	TCR (K^{-1})	References
Dc sputtering+oxidation	VO_x	673	2.0	Chen et al
PLD	VO_x	300	2.8	Rajendra Kumar et al
Ion beam sputtering+ oxidation	VO_2	473	2.6	Wang et al
RF sputtering	V_2O_5 /V/ V_2O_5	573	2.6	Moon et al
RF sputtering	V-W-O	573	2.6	Han et al
dc magnetron sputtering+annealing	VO_2	673	4.4	Yuqiang wt al
Reactive e-beam evaporation	$VO_2 + V_2O_5$	473	3.2	Subrahmanyam et al

Table 1. A summary of different deposition techniques for the growth of vanadium oxide with the process temperature and the reported TCR values.

A lot of efforts were being made to improve the quality of resistive VOx films and to obtain TCR values above 3%K^{-1}. The success was achieved by introducing tungsten-doping in a multilayer structure of V_2O_5 (Y. H. Han, S. H. Lee, K. T. Kim, I. H. Choi, & Moon, 2007; Y. Han, 2003; Moon, 2005). These oxide layers were deposited by reactive dc sputtering followed by an annealing treatment (673K). The analysis showed a TCR value of ~−4.4% °K^{-1} and a sheet resistance of 20 k Ω/square (Dai, X. Wang, He, Huang, & Yi, 2008; Lv et al., 2007). Although these results indicate a breakthrough for material performance, this material is not suitable for micro-machining process on Si and the fabrication of bolometers due to high temperature budget.

As discussed above, the $1/f$ noise is also an important figure of merit for thermistor materials. The noise in the oxide layers is caused mainly from the induced mechanical stresses due to the large grain sizes in the mixed phases. Zerov et al (Zerov, 2001) showed that the noise level in the oxide films is originated from two principal parameters: the concentration of different phases of VO_x and the grain size.

A recent study shows that high tungsten contents in vanadium oxide films (alloys of $V_{1-x}W_xO_2$ or VWO) will improve the thermal performance of the oxide material. Takami et al

(Takami, Kawatani, Kanki, & Tanaka, 2010) showed that the TCR performance of $V_{1-x}W_xO_2$ films grown on $Al_2O_3(0001)$ depends strongly on tungsten content. The tungsten level has been optimized and $V_{0.85}W_{0.15}O_2$ demonstrates a TCR value of 10%/K at room temperature. Moreover, the TCR behavior is found to be almost independent of layer thickness which is very beneficial for bolometric application.

3.2 Single-crystalline Si(C)/SiGe(C) multilayer structures

Many initiatives were taken to improve the IR detection and to obtain high quality imaging. Most of these efforts have striven to increase SNR and the thermal response of the detector. For bolometers, $1/f$ noise is the main source of noise (Lv et al., 2007).

Single crystalline semiconductor heterostructures are outstanding alternatives for low noise thermistor material. Among low cost semiconductor materials, SiGe(C)/Si(C) MQWs are the most appealing alternatives due to its low noise performance. This material system is therefore very promising for future mass-market applications. The structures demonstrate low noise when high quality of epi-layers, interfacial roughness (or unevenness) and the contact resistances are obtained (Kolahdouz, Afshar Farniya, Di Benedetto, et al., 2010; Kolahdouz, Afshar Farniya, Östling, & H. Radamson, 2010).

When the semiconductor thermistor material is heated, thermal excitations generate carriers (holes in this case) which have energies high enough to overcome the potential barrier of the quantum well. If a voltage is applied across the active region, these excited carriers move in the direction of the applied field, thus resulting in a current (see Fig. 5). This current increases at higher temperatures by increasing the number of the carriers in the current stream.

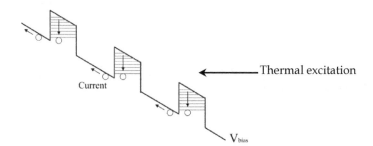

Fig. 5. When a voltage is applied across the thermistor, the valence band in SiGe/Si is tilted and thermally excited holes move towards the negative potential.

Kolahdouz et al. (Kolahdouz, Afshar Farniya, Di Benedetto, et al., 2010) presented the effect of Ge content (barrier height) on the performance of the SiGe/Si multi quantum wells (MQWs) and dots (MQDs) as thermistor material. In this study three Ge contents (23, 28 and 32%) in SiGe wells were applied and for higher Ge content (~47 %), Ge-dots/Si systems were grown. In order to have a decent growth rate, the samples were grown at 600 °C. At this growth temperature, the intermixing of Si into Ge makes it impossible to grow pure Ge dots. This problem makes these structures vulnerable to strain relaxation and defect formation.

The experimental data demonstrated a TCR value of ~3.4 %K⁻¹ for Ge MQDs which is a clear improvement compared to SiGe wells layers with 2.7 %K⁻¹. However, a remarkable increase of the noise constant ($k_{1/f}$) is observed for MQDs compared to MQWs (see Fig.6). It is believed that the noise level is sensitive to the variation of hole concentration in the Ge-dot systems' structures compared to the uniform profile in SiGe wells. Any strain relaxation in Ge dots will contribute to the noise level. A summary of both MQW and MQD SiGe/Si is presented in table 2.

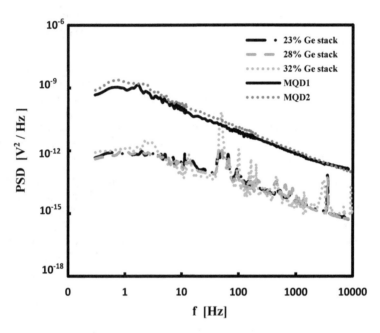

Fig. 6. Noise power spectral density of devices vs. frequency for pixel area 70×70 μm² in Si/SiGe MQWs and MQDs.

	Ref. Si	MQW1	MQW2	MQW3	MQD1	MQD2
Ge content (%)	0	23	28	32	47	>47
Estimated barrier height (eV)[a]	⋯	0.19	0.23	0.27	⋯	⋯
TCR (%/K) for 200×200 μm²	0.12	1.39	1.87	2.42	2.51	2.87
TCR (%/K) for 140×140 μm²	0.09	1.55	2.06	2.50	2.86	3.14
TCR (%/K) for 100×100 μm²	0.08	1.66	2.15	2.55	2.92	3.23
TCR (%/K) for 70×70 μm²	0.08	1.70	2.77	2.57	2.98	3.4
$K_{1/f}$	9×10^{-13}	1×10^{-12}	3×10^{-12}	4.4×10^{-12}	2×10^{-9}	2×10^{-9}
Energy quantized levels (meV)	⋯	9.6,38,54,84.1,84.2	10.1,40,62,89,94	10.4,41,68,92,101	⋯	⋯
Normalized resistance (Ω mm²) (R_0=R×area=$p \times l$)	0.2	1.17	3.46	2.39	2.91	4.88

Table 2. Summary of estimated barrier heights, TCR(%/K), K₁/f, R₀, and energy levels in QWs (at room temperature) for all sizes of detector structures. Due to partial strain relaxation, the barrier height of the quantum dots is not specified.

These results indicate that the performance of the thermistor material SiGe/Si MQWs (or MQDs) is very sensitive to structure profile.

Andersson et al, (J. Y. Andersson, P. Ericsson, H. H. Radamson, S. G. E. Wissmar, & Kolahdouz, 2011) presented theoretical calculations to optimize TCR in terms of the Ge content and quantum well width in SiGe/Si MQW and MQD systems (see Fig.7). The results were also compared to the experimental data. The extracted TCR values for MQW structures showed that the thermal response of detectors increases with Ge content which is consistent with the experimental data. The authors propose Ge dots with high Ge content as a better solution for SiGe wells. The valence band off-set and TCR values versus the size of Ge dots were calculated (see Fig.8a). The ground level energy relative valence band edge in Si versus thequantum dots demonstrates the dependency of TCR on the size of Ge dots (see Fig.8b). The results show that dots with 60 nm could exhibit a TCR value of 8.5% which is excellent for IR detection. However, these calculations do not consider the noise level in MQD system. The growth of pure Ge dot on the Si surface is a challenging task due to the intermixing of Si into Ge. In order to avoid this problem, low temperature epitaxy can be applied to grow Ge dots with high Ge content in MQD structures. This low temperature process suffers from low growth rate and thus makes it impractical for mass production.

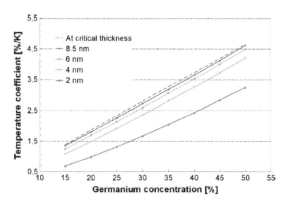

Fig. 7. The temperature coefficient versus Ge content in the QWs, for different QW widths. simulated data.

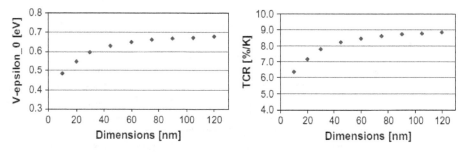

Fig. 8. a. Plots of ground level energy relative valence band edge in Si for different Ge dot sizes, and b. the dependence of the temperature coefficient of resistivity (TCR) on dot size.

Wissmar et al. (S. Wissmar et al., 2006) investigate the TCR performance of SiGe/Si and AlGaAs/GaAs sytems. This study also demonstrates the dependence of thermal performance of MQW structures versus the quantum well profile (composition, dopant concentration and the width of the quantum wells). The performance of the structures is degraded with increasing the dopant concentration in the quantum well. The AlGaAs/GaAs system demonstrates excellent performance (4.5%/K) compared to SiGe/Si system (2 %/K). However, the low cost Si technology is always preferred over III-V for industrial applications.

More discussions about the SiGe/Si thermistor material are presented by Ericsson et al. (Per Ericsson et al., 2010) It is generally observed that the flicker noise is volume dependent (Motchenbacher, 1973). This study presents the effect of pixel area on the Flicker noise (the vertical thickness is constant). The data show the dependency of 1/A as expected by theories (see Fig. 9).

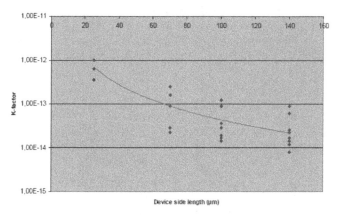

Fig. 9. The flicker noise k-value for SiGe/Si MQW structures with different pixel sizes and the predicted variation (solid line). All data have been manufactured on the same wafer.

A vital issue in many cases for bolometers containing SiGe/Si MQW system is the control of the residual strain in the suspended membrane.

Radamson et al. (H. H. Radamson, Kolahdouz, Shayestehaminzadeh, Afshar Farniya, & S. Wissmar, 2010) reports the integration of C in the Si/SiGe stack (SiGe(C) / Si(C) MQWs) to create alternating tensile/compressive strain systems. The SiGe(C) layers were created through the intermixing of Si into the embedded Ge thin layers (grown by introducing GeH_4 without SiH_4). The intermixing of Si and Ge can be controlled by the growth temperature and the carbon doping in the Si barrier layer (Hirano & Murota, 2009). This study compares five different structure profiles considering the effect of contact resistance (Ni silicide contacts), Ge content, and carbon doping in Si barrier (see Table 3). The prototypes exhibited an outstanding TCR of 4.5%/K for $100 \times 100 \mu m^2$ pixel sizes and low noise constant ($K_{1/f}$) value of 4.4×10^{-15}. The excellent performance of the devices was due to low contact resistance in presence of Ni silicide contacts, smooth interfaces, and high quality multi quantum wells (MQWs) containing high Ge content. Fig.10 demonstrates the noise data for the samples described in table 3. Samples MQW1 (no silicide contacts) and MQW5 ($Si_{0.35}Ge_{0.65}$/SiC with silicide contacts) show the highest and the lowest noise level among this sample series.

	MQW1	MQW2 (silicide)	MQW3 (C in barrier)	MQW4 (Ge-delta+silicide)	MQW5 (Ge-delta+ silicide+C in barrier)
Ge content[%]	28	28	>28	10	65
Estimated barrier height (eV)	0.23	0.23	0.23	0.07	0.5
Activation energy (eV) for 100×100µm²	0.17	0.18	0.19	0.07	0.35
TCR [%/K] for 200×200µm²	1.87	2.32	2.75	0.6	4
TCR [%/K] for 100×100µm²	2.77	2.30	2.78	1.07	4.5
$K_{1/f}$ for 100×100µm²	$4.1×10^{-12}$	$1.4×10^{-13}$	$5×10^{-14}$	$8.5×10^{-15}$	$4.4×10^{-15}$
Normalized Resistance [Ωmm²] $(R_0 = R×area = \rho×l)$	3.46	2.67	19.66	0.4	3.11

Table 3. Summary of estimated barrier heights, TCR [%/K], K1/f and R0 in MQWs (at room temperature) for all sizes of detector.

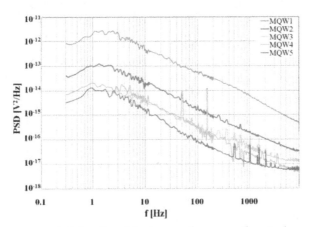

Fig. 10. Noise power spectral density of devices vs. frequency for pixel area 100×100 µm² in SiGe MQWs described in table 3.

4. Fabrication process flow

Bolometers are mainly composed of a temperature sensing resistor and an IR absorber. A good thermal isolation is the requirement to increase the sensitivity of these detectors. This can be achieved by suspending the bolometer structure in the air through either membrane or bridge support as shown in Fig. 11.

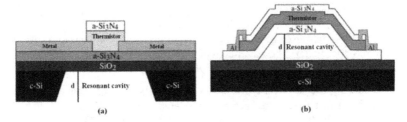

Fig. 11. Cross-section of a) the membrane-supported and b) the bridge-supported microbolometer (Garcia, 2004).

It was reported in 2004 (Garcia, 2004) that the noise current of the bridge–supported structures is one order of magnitude higher than that of the membrane–supported structure. However, the bridge–supported structure process flow enables a precise control on the resonant cavity length which makes it the dominant design for microbolometers.

The process flow of fabricating a bolometer is very dependent on the thermistor material. For thermistors which may be deposited at low temperatures, there is a possibility of being directly integrated on the readout integrated circuit (ROIC) without harming its elements. Amorphous Si, SiGe, Ge, $Ge_xSi_{1-x}O_y$ and poly VOx material are a few examples of such thermistors. The advantage of the mentioned group IV–based materials in this list is their absolute compatibility with the silicon processing line.

The process flow is described in Fig. 12 (Mottin et al., 2002). It is the simplest manufacturing method in which the thermistor material can be grown directly on ROIC. The first step is the deposition of a thin reflective layer directly on top of the ROIC. A thick sacrificial layer is then spun and cured to form the resonant cavity at the end of the process. The thermistor material is deposited over the sacrificial layer and covered by the metallic contact electrodes. The metallic contact deposition and etching enable electrical continuity between the underlying substrate and the thermistor on the surface of the sacrificial layer. Finally, the micro–bridge arrays are released by removing the sacrificial layer.

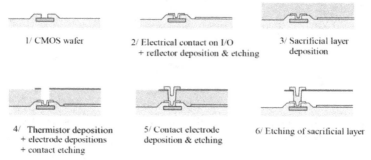

Fig. 12. Process flow of a bridge–supported microbolometer technology (Mottin et al., 2002).

The second fabrication method is based on wafer bonding where thermistor material is transferred from epi-wafer to RIOC wafer. This is necessary since a thermal treatment (850-900 C) is required for in-situ cleaning prior to epitaxy of single-crystalline layers. A process flow for fabrication of bolometers based on structures composed of sc- group IV materials through wafer bonding process on ROICs (Kvisterøy et al., 2007; J. Källhammer et al., 2006; F. Niklaus et al., 2001) is demonstrated in Fig. 13. In this process, group IV-based structures are deposited on a separate SOI carrier wafer and are then transferred from the handle wafer to the ROIC wafer using low-temperature adhesive wafer bonding in combination with sacrificial removing of the carrier wafer. The advantage of 3D bolometer integration is that it allows the employment of high TCR and SNR mono-crystalline thermistor for imaging applications.

For many detector applications, Ni is chosen as the absorbent layer. This is due to its simple preparation and the fact that it can get a strong absorption of about 90% in wavelength range between 7-13 μm (Lienhard, Heepmann, & Ploss, 1995). In 2006 Hsieh et al. (Hsieh, Fang, & Jair, 2006) reported TCR value of -2.74 %K[-1] and activation energy of 0.21eV for $Si_{0.68}Ge_{0.31}C_{0.01}$ ternary system.

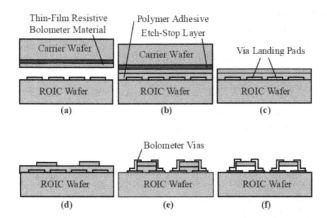

Fig. 13. A Schematic picture of a Si-based bolometer process (F. Niklaus et al., 2007).

The sc-SiGe/Si structure is transferred to the ROIC by low temperature adhesive wafer bonding and subsequent removal of the carrier. In 2010, Lapadatu et al. (Lapadatu et al., 2010) proposed a novel approach to increase the fill factor. In their design the legs, which support the bolometer membrane and connect it to the ROIC, are built underneath the membrane as shown in Fig. 14.

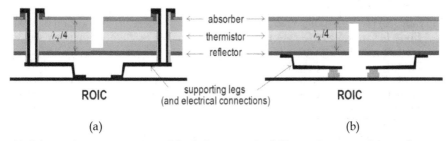

(a)   (b)

Fig. 14. Schematic representation of the bolometer pixel illustrating two schemes for electrical connection (Lapadatu et al., 2010) (a) through-pixel plugs; (b) under-pixel plugs.

It was reported that the detectors composed of SiGe quantum wells have presented a TCR around 3.1 %K^{-1} and 5×10^{-13} for K$_{1/f}$ (Lapadatu et al., 2010).

It is important to emphasize here that the recent advanced cleaning technique together with new gas precursor for Si (trisilane) and Ge (digermane) may provide the opportunity to grow epi-layers at low temperatures (300-500 °C). This means that the fabrication technique will become similar to the steps in Fig.12 and sc-Si-based material will be deposited directly on ROIC wafer.

5. Conclusions

Among different materials, single crystalline SiGe alloy is a promising thermistor material in bolometers for LWIR detections. The temperature response of SiGe/Si multi quantum well (or dot) structures depends mainly on Ge content (strain). The signal-to-noise ratio which is an important parameter for thermal imaging is strongly sensitive to contact resistance, interfacial and layer quality of SiGe layers. It is demonstrated that carbon-doping impedes the defect formation in SiGe layers and SiGe(C)/SiC with Ni silicide contacts is the most attractive structure for high performance bolometers.

6. References

Andersson, J. Y., Ericsson, P., Radamson, H. H., Wissmar, S. G. E., & Kolahdouz, M. (2011). SiGe/Si quantum structures as a thermistor material for low cost IR microbolometer focal plane arrays. *Solid-State Electronics*, *60*(1), 100-104. Elsevier Ltd. doi:10.1016/j.sse.2011.01.034

Bean, J. C., Feldman, L. C., Fiory, A. T., Nakahara, S., & Robinson, I. K. (1984). Gex Si1 _ x lSi strained-layer superlattice grown by molecular beam epitaxy. *J. Vac. Sci. Techno. A*, *2*(2), 436-440.

Di Benedetto, L., Kolahdouz, M., Malm, B. G., Ostling, M., & Radamson, H. H. (2009). Strain balance approach for optimized signal-to-noise ratio in SiGe quantum well bolometers. *ESSDERC 2009 - Proceedings of the 39th European Solid-State Device Research Conference* (pp. 101-104).

Cabarcos, O. M., Li, J., Gauntt, B. D., Antrazi, S., Dickey, E. C., Allara, D. L., & Horn, M. W. (2011). Comparison of ion beam and magnetron sputtered vanadium oxide thin films for uncooled IR imaging. *Proc. of SPIE Vol. 8012* (Vol. 8012, p. 80121K-80121K-9). doi:10.1117/12.884377

Cohen-Tannoudji, C., Diu, B., & Laloe, F. (1977). *Quantum Mechanics*. New York, USA: John Wiley and Sons.

Dai, J., Wang, X., He, S., Huang, Y., & Yi, X. (2008). Low temperature fabrication of VOx thin films for uncooled IR detectors by direct current reactive magnetron sputtering method. *Infrared Physics & Technology*, *51*(4), 287-291. doi:10.1016/j.infrared.2007.12.002

Ericsson, Per, Höglund, Linda, Samel, B., Savage, Susan, Wissmar, Stanley, Oberg, O., Kallhammer, J.-E., et al. (2010). Design and evaluation of a quantum-well-based resistive far-infrared bolometer. *Proc. of SPIE Vol. 7834* (Vol. 7834, pp. 78340Q1-10). doi:10.1117/12.865036

Garcia, M. (2004). IR bolometers based on amorphous silicon germanium alloys. *Journal of Non-Crystalline Solids*, *338-340*, 744-748. doi:10.1016/j.jnoncrysol.2004.03.082

Han, Y. H., Lee, S. H., Kim, K. T., Choi, I. H., & Moon, S. (2007). Properties of electrical conductivity of amorphous tungsten-doped vanadium oxide for uncooled microbolometers. *Solid State Phenom.*, *343*, 124-126.

Han, Y., Choi, I., Kang, H., Parkb, J., Kimb, K., Shin, H., & Moon, S. (2003). Fabrication of vanadium oxide thin film with high-temperature coefficient of resistance using V2O5/V/V2O5 multi-layers for uncooled microbolometers. *Thin Solid Films*, *425*(1-2), 260-264. doi:10.1016/S0040-6090(02)01263-4

Hirano, T., & Murota, J. (2009). A Study on Formation of Strain Introduced Group IV Semiconductor Heterostructures by Atomic Layer Doping. *Record of Electrical and Communication Engineering Conversazione Tohoku University*, *78*(1), 407-8.

Hooge, F. N. (1994). 1/F Noise Sources. *IEEE Transactions on Electron Devices*, *41*(11), 1926-1935. doi:10.1109/16.333808

Hsieh, M., Fang, Y. K., & Jair, D. K. (2006). The study of a novel crystal SiGeC far infrared sensor with thermal isolated by MEMS technology. *Microsystem Technologies*, *12*(10-11), 999-1004. doi:10.1007/s00542-006-0162-7

Kolahdouz, M., Afshar Farniya, A., Di Benedetto, L., & Radamson, H. (2010). Improvement of infrared detection using Ge quantum dots multilayer structure. *Applied Physics Letters*, *96*(21), 213516. doi:10.1063/1.3441120

Kolahdouz, M., Afshar Farniya, A., Östling, M., & Radamson, H. (2010). Improving the performance of SiGe-based IR detectors. *ECS Transactions, 33 (6)* (Vol. 33, pp. 221-225).

Kruse, P. W., McGlauchlin, L. D., & McQuistan, R. B. (1962). *Elements of Infrared Technology. Generation, transmission, and detection. Wiley, New York,* (Vol. 137, pp. 123-123). Wiley, New York,. doi:10.1126/science.137.3524.123

Kumar, R. T. R., Karunagaran, B., Mangalaraj, D., Narayandass, S. K., Manoravi, P., Joseph, M., & Gopal, V. (2003). Study of a pulsed laser deposited vanadium oxide based microbolometer array. *Smart Materials and Structures*, *12*(2), 188-192. doi:10.1088/0964-1726/12/2/305

Kvisterøy, T., Jakobsen, H., Vieider, C., Wissmar, S., Ericsson, P., Halldin, U., Niklaus, Frank, et al. (2007). Far infrared low-cost uncooled bolometer for automotive use. *Proc. AMAA.* Berlin, Germany.

Källhammer, J., Pettersson, H., Eriksson, D., Junique, S., Savage, S., Vieider, C., Andersson, J. Y., et al. (2006). Fulfilling the pedestrian protection directive using a long-wavelength infrared camera designed to meet both performance and cost targets. *Proceedings of SPIE, 6198*, 619809-1. Spie. doi:10.1117/12.663152

Lapadatu, A., Kittilsland, G., Elfving, A., Hohler, E., Kvisterøy, T., Bakke, T., & Ericsson, P. (2010). High-performance long wave infrared bolometer fabricated by wafer bonding. *Technology, 7660*, 766016-766016-12. doi:10.1117/12.852526

Liddiard, K. C. (1984). Thin-film resistance bolometer IR detectors. *Infrared Physics*, *24*(1), 57-64. doi:10.1016/0020-0891(84)90048-4

Lienhard, D., Heepmann, F., & Ploss, B. (1995). Thin nickel films as absorbers in pyroelectric sensor arrays. *Microelectronic Engineering*, *29*, 101-104.

Lv, Y., Hu, M., Wu, M., & Liu, Z. (2007). Preparation of vanadium oxide thin films with high temperature coefficient of resistance by facing targets d.c. reactive sputtering and annealing process. *Surface and Coatings Technology*, *201*(9-11), 4969-4972. doi:10.1016/j.surfcoat.2006.07.211

Malyarov, V. G., Khrebtov, I. A., Kulikov, Y. V., Shaganov, I. I., Zerov, V. Y., & Vavilov, S. I. (1999). Comparative investigations of the bolometric properties of thin film structures based on vanadium dioxide and amorphous hydrated silicon. *Proceedings of SPIE 3819* (Vol. 3819, pp. 136-142). Spie. doi:10.1117/12.350896

Mathur, S., Ruegamer, T., & Grobelsek, I. (2007). Phase-Selective CVD of Vanadium Oxide Nanostructures. *Chemical Vapor Deposition*, *13*(1), 42-47. doi:10.1002/cvde.200606578

Moon, S., Han, Y., Kim, K., Lee, S., & Shin, H. (2005). Enhanced Characteristics of V0.95W0.05OX-Based Uncooled Microbolometer. *IEEE Sensors, 2005.*, 1137-1140. Ieee. doi:10.1109/ICSENS.2005.1597905

Moreno, M., Kosarev, A., Torres, A., & Ambrosio, R. (2007). Fabrication and performance comparison of planar and sandwich structures of micro-bolometers with Ge thermo-sensing layer. *Current, 515*, 7607 - 7610. doi:10.1016/j.tsf.2006.11.172

Motchenbacher, C. D. (1973). *Low Noise Electronic Design.* New York, USA: John Wiley & Sons, Inc.

Mottin, E., Bain, A., Martin, J. L., Ouvrier-Buffet, J. L., YonN, J. J., Chatard, J. P., & Tissot, J. L. (2002). Uncooled amorphous-silicon technology: high-performance achievement and future trends. *Proceedings of SPIE, 4721*, 56-63. Spie. doi:10.1117/12.478861

Niklaus, F., Kälvesten, E., & Stemme, G. (2001). Wafer-level membrane transfer bonding of polycrystalline silicon bolometers for use in infrared focal plane arrays. *Journal of Micromechanics and Microengineering*, 11, 509-513.

Niklaus, F., Vieider, C., & Jakobsen, H. (2007). MEMS-based uncooled infrared bolometer arrays: a review. *Proceedings of SPIE*, 6836, 68360D-68360D-15. Spie. doi:10.1117/12.755128

Niklaus, Frank, Decharat, A., Jansson, C., & Stemme, Göran. (2008). Performance model for uncooled infrared bolometer arrays and performance predictions of bolometers operating at atmospheric pressure. *Infrared Physics & Technology*, 51(3), 168-177. doi:10.1016/j.infrared.2007.08.001

Radamson, H. H., Kolahdouz, M., Shayestehaminzadeh, S., Afshar Farniya, A., & Wissmar, S. (2010). Carbon-doped single-crystalline SiGe / Si thermistor with high temperature coefficient of resistance and low noise level. *Applied Physics Letters*, 97(23), 223507. doi:L10-06859R

Rogalski, a. (2011). Recent progress in infrared detector technologies. *Infrared Physics & Technology*, 54(3), 136-154. Elsevier B.V. doi:10.1016/j.infrared.2010.12.003

Schimert, T., Brady, J., Fagan, T., Taylor, M., McCardel, W., Gooch, R., Ajmera, S., et al. (2008). Amorphous silicon based large format uncooled FPA microbolometer technology. *Proceedings of SPIE*, 6940, 694023-694023-7. Spie. doi:10.1117/12.784661

Smith, A., Jones, F. E., & Chasmar, R. P. (1968). *The Detection and Measurement of Infrared Radiation*. Clarendon, Oxford,.

Sood, A. K., Richwine, R. a, Puri, Y. R., DiLello, N., Hoyt, J. L., Akinwande, T. I., Dhar, N., et al. (2010). Development of low dark current SiGe-detector arrays for visible-NIR imaging sensor. *Proc. of SPIE Vol. 7660* (Vol. 7660, pp. 76600L1-7). doi:10.1117/12.852682

Subrahmanyam, A., Bharat Kumar Redd, Y., & Nagendra, C. L. (2008). Nano-vanadium oxide thin films in mixed phase for microbolometer applications. *Journal of Physics D: Applied Physics*, 41(19), 195108. doi:10.1088/0022-3727/41/19/195108

Takami, H., Kawatani, K., Kanki, T., & Tanaka, H. (2010). High Temperature-Coefficient of Resistance at Room Temperature in W-Doped VO2 Thin Films on Al2O3 Substrate and Their Thickness Dependence. *Jpn. J. Appl. Phys.*, 50, 055804.

Vieider, C., Wissmar, S., Ericsson, P., Halldin, U., Niklaus, Frank, Stemme, G., Kallhammer, J., et al. (2007). Low-cost far infrared bolometer camera for automotive use. *Proceedings of SPIE*, 6542, 65421L-65421L-10. Spie. doi:10.1117/12.721272

Wang, H., Yi, X., & Chen, S. (2006). Low temperature fabrication of vanadium oxide films for uncooled bolometric detectors. *Infrared Physics & Technology*, 47(3), 273-277. doi:10.1016/j.infrared.2005.04.001

Wissmar, S., Höglund, L., Andersson, J., Vieider, C., Savage, S., & Ericsson, P. (2006). High signal-to-noise ratio quantum well bolometer materials. *Proceedings of SPIE*, 6401, 64010N-64010N-11. Spie. doi:10.1117/12.689874

Wissmar, S., Radamson, H., Kolahdouz, M., & Andersson, J. Y. (2008). Ge quantum dots on silicon for terahertz detection. *TERA* (Vol. 6542, pp. 6542-6542).

Zerov, V. Y., Kulikov, Y. V., Malyarov, V. G., Khrebtov, I. a, Shaganov, I. I., & Shadrin, E. B. (2001). Vanadium oxide films with improved characteristics for ir microbolometric matrices. *Technical Physics Letters*, 27(5), 378-380. doi:10.1134/1.1376757

Part 2

Bolometer Types and Properties

3

Cold-Electron Bolometer

Leonid S. Kuzmin

Chalmers University of Technology,
Department of Microtechnology and Nanoscience
Sweden

1. Introduction

Cosmic microwave background (CMB) measurements are ranked second by the journal *Science* among the top 10 Achievements of the Decade. In 2000 and 2003, an experiment known as balloon observations of millimetric extragalactic radiation and geophysics (BOOMERanG) measured the CMB in detail in patches of the sky. Then in 2003, NASA's space-based Wilkinson microwave anisotropy probe (WMAP) mapped the CMB across the sky, producing an exquisite miniature picture of the cosmos. These and the measurements that followed have started transforming cosmology from a largely qualitative endeavour to a precision science with a standard theory, named 'precision cosmology' (Cho, 2010; Masi, 2006). Recent cosmology experiments have discovered that the universe consists mainly of dark energy and dark matter.

Experiments to resolve the nature of these dark components will require a new generation of ultra-sensitive detectors. At present, the most widespread superconducting bolometer is a transition-edge sensor (TES) (Irvin, 1995; Lee et al., 1996). However, due to artificial dc bias heating, the TES has excess noise and strictly limited saturation power.

A novel concept of the **Cold-Electron Bolometer (CEB)** (Kuzmin, 1998, 2000, 2001, 2004; Kuzmin & Golubev, 2002) has been invented to overcome these problems. The CEB concept is based on a unique combination of the *RF capacitive coupling* of an absorber to the antenna through capacitance of the SIN tunnel junctions (Kuzmin, 1998, 2000) and *direct electron cooling* of the absorber by the same SIN tunnel junction (Kuzmin, 2001; Kuzmin & Golubev, 2002). The noise properties of this device are improved considerably by decreasing the electron temperature. Direct electron cooling leads also to a considerable increase of the saturation power due to removing incoming power from the sensitive nanoabsorber. Direct electron cooling provides strong *negative electrothermal feedback* for the signal (Kuzmin, 2004), analogous to the TES (Irvin, 1995). However, the *artificial dc heating* for feedback in TES is replaced by using an effect of electron cooling by SIN tunnel junctions (Nahum 1994) for *direct electron cooling* of the absorber by SIN tunnel junctions (Kuzmin 1998; Golubev & Kuzmin, 2001) to a minimum temperature (that could be less than phonon temperature). This concept can lead to a major breakthrough in the realisation of supersensitive detectors.

Historically, development of SIN tunnel junctions as a thermometer was started in the 1970s (Bakker et al., 1970). The SIN junction demonstrated sensitivity to the temperature of a normal metal. A normal metal hot-electron bolometer (NHEB) with SN Andreev mirrors for thermal isolation and a SIN tunnel junction for readout was proposed in 1993 (Nahum et al., 1993) and realized in the same year (Nahum & Martinis, 1993). Disadvantages of the NHEB are:

- quite a small dynamic range due to low saturation power for the ultrasensitive bolometer with a small absorber (SIN junction working as a voltmeter)
- poisoning of the bolometer by quasiparticles due to entrapping through Andreev contacts (Kuzmin et al., 1999)
- very complicated RF structure is disturbed by a central electrode for dc readout; in particular, it is impossible to create an array of bolometers (the most advanced concept of CEBs for the moment).

The effect of electron microrefrigeration by SIN tunnel junction was demonstrated in 1994 (Nahum et al., 1994). Quite effective cooling of the small absorber from 300 to 100 mK was demonstrated in several groups (Leivo et al., 1996; Kuzmin et al., 2004). Then several groups moved in the direction of using this microrefrigeration for cooling the platform with the detectors placed on this platform. However, this direction did not show good progress because of dramatic difficulties in cooling at low temperatures due to weak electron-phonon interaction and the fact that it only influenced the electron-phonon noise component. The main component, background noise, was not influenced by electron cooling.

A decisive step in the development of superconducting detectors has been the invention of a transition-edge sensor (TES) with strong electrothermal feedback (Irvin, 1995; Lee et al., 1996). However, the TES has some problems with excess noise, saturation and the most dramatic problem of artificial overheating by dc power for the electrothermal feedback. This unavoidable additional heating kills all efforts on deep cooling and does not give good prospects for the realisation of ultimate performance of the bolometer.

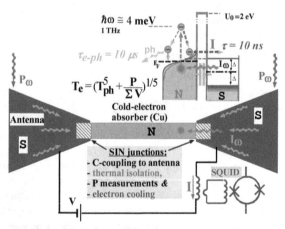

Fig. 1. Capacitively coupled cold-electron bolometer with nanoabsorber and SIN tunnel junctions for direct electron cooling and power measurements. The signal power is supplied to the sensor through capacitance of tunnel junctions, dissipated in the nanoabsorber and removed back from the absorber as hot electrons by the same SIN junctions. The electron cooling serves as strong negative electrothermal feedback improving all characteristics of the CEB: time constant, responsivity and NEP.

In contrast to this overheating, a new concept of a CEB with direct electron cooling has been proposed by Kuzmin (Kuzmin, 1998, 2000, 2001, 2004; Kuzmin & Golubev, 2002). The CEB is the only concept removing incoming background power from the supersensitive region of absorber. The CEB avoids the main problem of TES, an *additional dc heating* for the electrothermal feedback, by replacing it with *the direct electron cooling* of the absorber that could be a *turning point* in the realisation of modern supersensitive detectors. This cooling could be especially important for the realisation of high sensitivity in the presence of the realistic background power load because it returns the system to lowest temperature (noise) state. It could help to avoid full saturation when the signal exceeds the level of dc bias power that is the great problem for the TES. The CEB could give a high dynamic range in combination with the SQUID readout system having a high dynamic range in closed-loop operation. In this state the system shows the most reaction (responsivity) to an incoming signal. All power of the signal is used for measurements. A possible objection that tunnelling of electrons would increase shot noise is rejected by simple argument: if power is not removed by tunnel junctions, the same type of shot noise will be created by phonons through increased electron-phonon interaction.

2. Main concepts of the Cold-Electron Bolometer (CEB)

The basic concept of the CEB is a *cold-electron bolometer with capacitive coupling* to the antenna by two SIN tunnel junctions (Kuzmin, 2000, 2001; Kuzmin & Golubev, 2002). Theoretical estimations and preliminary experiments show that it is possible to realize the necessary sensitivity of better than 10^{-18} W/Hz$^{1/2}$ with antenna-coupled nanobolometers at a temperature of ≤0.3 K. Additional advantages of such detectors are the possibility to operate in a wide range of background load, easy integration in arrays on planar Si substrate and the possibility of polarisation measurements. Flexibility of the CEB concept gives the opportunity to realise this bolometer for any power load from 0.02 fW for space applications to 10 pW for ground-based applications with NEP less than photon noise of the signal.

To match the CEB with the requested JFET readout, two novel concepts of CEB have been proposed: parallel/series array of CEBs and 2D focal plane array of CEBs with SIN tunnel junctions. Simulations show high performance of these bolometers with NEP less than photon noise of the signal. These concepts should be qualified experimentally in RF tests and implemented in BOOMERanG. These concepts could also be used in OLIMPO, SPICA, B-Pol and Millimetron cosmology instruments.

2.1 General theory of the CEB

The operation of CEB can be analysed using a heat balance equation (Golubev & Kuzmin, 2001; Kuzmin & Golubev, 2002)

$$P_{SIN}(V,T_e,T_{ph}) + \Sigma\Lambda(T_e^5 - T_{ph}^5) + C_\Lambda \frac{dT}{dt} = P_0 + \delta P(t) \tag{1}$$

Here, $\Sigma\Lambda(T_e^5 - T_{ph}^5)$ is the heat flow from electron to the phonon subsystems in the normal metal, Σ is a material constant, Λ - a volume of the absorber, T_e and T_{ph} are, respectively, the electron and phonon temperatures of the absorber; $P_{SIN}(V,T_e,T_{ph})$ is cooling power of the SIN tunnel junctions; $C_v = \gamma T_e$ is the specific heat capacity of the normal metal and $P(t)$ is the incoming RF power. We can separate Eq. (1) into the time independent equation, $\Sigma\Lambda(T_{e0}^5 - T_{ph}^5) + P_{SIN0}(V, T_{e0}, T_{ph}) = P_0$, and the time dependent equation,

$$(\frac{\partial P_{SIN}}{\partial T} + 5\Sigma\Lambda T_e^4 + i\omega C_\Lambda)\delta T = \delta P \tag{2}$$

The first term, $G_{SIN} = \partial P_{SIN}/\partial T$, is the cooling thermal conductance of the SIN junction that gives the negative electrothermal feedback (ETF); when it is large, it reduces the temperature response δT because cooling power, P_{SIN}, compensates the change of signal power in the bolometer. The second, $G_{e-ph} = 5\Sigma\Lambda T_e^4$, is electron-phonon thermal conductance of the absorber. From Eq. (2) we define an effective complex thermal conductance which controls the temperature response of CEB to the incident signal power

$$G_{eff} = G_{cool} + G_{e-ph} + i\omega C_\Lambda \tag{3}$$

Full expression for current is

$$I = \frac{1}{eR} \int dE N_S(E)[f(E, Te) - f(E - eV, Tph)] + \frac{V}{Rj} \tag{4}$$

where $N_S(E)= |E|/\sqrt{(E^2 - \Delta^2)}$ is the normalized density of states in the superconductor and $f(E,T)=1/[\exp(E/T)+1]$.

In analogy with TES (Irvin, 1995; Lee et al.,1996), the effective thermal conductance of the CEB is increased by the effect of electron cooling (negative ETF).

The current responsivity is given by

$$S_i = \frac{\partial I}{\partial P} = \frac{\partial I / \partial T}{G_{cool} + G_{e-ph} + i\omega C_\Lambda} = \frac{\partial I / \partial T}{G_{cool}} \frac{L}{(L+1)[1+i\omega\tau]}, \tag{5}$$

where $L = G_{cool}/G_{e-ph} \gg 1$ is ETF gain and

$$\tau = C_\Lambda/G_{e-ph} = \tau_0/(L+1) \tag{6}$$

is an effective time constant, $\tau_0 = C_\Lambda/G_{e-ph}$ ($\cong 10\mu s$ at 100 mK).

Strength of electrothermal feedback is estimated as:

$$L(\omega) = \frac{G_{cool}}{G_{e-ph}(1+i\omega\tau)} = \frac{\partial I / \partial T}{G_{cool} + G_{e-ph} + i\omega C_\Lambda} \tag{7}$$

2.1.1 CEB and TES comparison

We compare now the realization of CEB and TES. The principle of operation is shown in Fig. 3. Both concepts use the voltage-biased mode of operation. The TES is heated to critical temperature by dc power P_{bias} (Fig. 2a). This temperature is supported during all range of operation (before saturation). Electrothermal feedback arises from the dependence of the bias power on the resistance of the superconductor. If there is an increase in optical power incident on the bolometer, the bias power decreases and nearly compensates for the increase incident power (Fig. 2b). Output signal is this decrease in bias power nearly equals the incident power.

The principle of operation of the CEB (Fig. 2a) is approximately the same, but moving to another bias point in temperature: to absolute zero. Starting from the phonon temperature T_{ph}=100 mK, the cooling conductance, G_{cool}, decrease the electron temperature to the possible minimum level, 30 mK in this case (for typical parameters of CEB: Λ=0.01 μm^3 and R=1 kΩ). After applying signal power, the cooling conductance increases trying to compensate for the increase in the electron temperature to the minimum temperature close to the previous value.

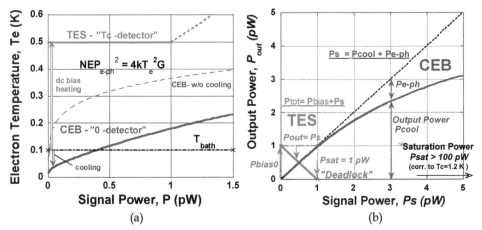

Fig. 2. (a) Electron temperature as a function of signal power for T_{bath}=100 mK for CEB and TES. For CEB, the T_e is always cooled to the lowest possible level. For P<0.4pW, the T_e of CEB is less than T_{bath} (real cold-electron bolometer). For TES, the T_e is equal to T_C for all range of operation up to saturation power. After saturation there is uncontrollable increase of temperature. (b) Output power (cooling power) of the CEB in dependence on signal power (blue). They are almost equal at a small level of power. At a higher level of power, the P is split between Pcool and Pe-ph. The saturation power would be achieved only after heating to Tc of Al electrode (P_{sat} is around 100 pW). Bias power and output power of the TES on signal power (red). Saturation power is equal to the full bias power at zero signal.

Dependence of output power on signal power is shown in Fig. 2b. For both concepts, Pout is nearly equal to incoming power in the range of dc heating power (TES) and typical cooling power (CEB). Accuracy of removing (CEB) or compensation (TES) of incoming power is determined by the strength of the electrothermal feedback - loop gain L. For TES, the L is determined by nonlinearity of R(T) dependence and could exceed 1000. For CEB, the L is determined as relation of thermal conductances (6). Typical dependence of L on incoming power is shown in Fig. 3.

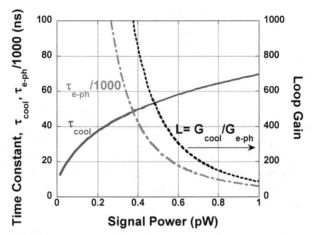

Fig. 3. Time constant of CEB, τ_{cool}, in dependence on signal power P_s. The electron-phonon time constant, τ_{e-ph}, is shown for comparison (scaled 1000 times). The τ_{cool} is considerably shorter than τ_{e-ph} and difference is increased when we move to smaller signal power with stronger electron cooling.

The saturation problem is very serious for TES: P_{sat} is exactly equal to applied dc heating power P_{bias} (Fig. 2b). If we increase saturation level, the overheating of the TES would unavoidably be increased. After saturation power, the TES stops operation fully. It is difficult to foresee the expected level of maximum power load and choosing P_{sat} is a really complicated problem.

The situation for CEB saturation is completely different. There is no dramatic problem at all: the output cooling power would simply deviate from the linear dependence Pcool(P). For the typical cooling power around 1 pW, the deviation from linear dependence $P_{out}(P)$ would be only several % at this level (Fig. 2b). As signal power is further increased, the deviation will be larger, but the CEB still continues to work. It is only a question of calibration of this dependence. Final 'deadlock' for the CEB would be at the level of power around 100 pW when temperature achieves the critical temperature of the Al electrode. Thus, due to absence of artificial heating, the CEB does not really have the saturation problem and can considerably extend possible scope of operation.

2.1.2 Time constant

The time response of the CEB (6) in dependence on incoming power is shown in Fig. 3. As for TES, it is strongly reduced by loop gain L (7) of electrothermal feedback. Cooling conductance G_{cool} is not dependent strongly on incoming power and slightly reduced for smaller power. In contrast, the e-ph conductance , G_{e-ph}, is very dependent on power due to 4th dependence on electron temperature and strongly increased for low power as well as related e-ph time constant. As a final result, the L is considerably increased for small power and exceeds the level of 1000.

The main characteristic of negative electrothermal feedback, the loop gain L, is shown by the dashed line. The loop gain is strongly increased for smaller P_s due completely to the decrease of G_{e-ph}.

2.1.3 Noise Properties (NEP)

Noise properties are characterized by the noise equivalent power (*NEP*), which is the sum of three different contributions and is defined as follows:

$$NEP_{total}^2 = NEP_{e-ph}^2 + NEP_{SIN}^2 + \frac{\delta I^2}{S_I^2}. \tag{8}$$

Here

$$NEP_{e-ph}^2 = 10k_B \Sigma \Lambda (T_e^6 + T_{ph}^6) \tag{9}$$

is the noise associated with electron-phonon interaction (Golwala, S. et al., 1997; Golubev & Kuzmin, 2001) NEP^2_{SIN} is the noise of the NIS tunnel junctions and the last term $\delta I^2/S^2_I$ is due to the finite sensitivity of the amplifier (SQUID) δI, which is expressed in pA/Hz$^{1/2}$.

The noise of the SIN tunnel junctions, NEP^2_{SIN}, has three components: shot noise $2eI/S^2_I$, the fluctuations of the heat flow through the tunnel junctions and the correlation between these two processes

$$NEP_{SIN}^2 = \frac{\delta I_\omega^2}{S_I^2} - 2\frac{<\delta P_\omega \delta I_\omega>}{S_I} + \delta P_\omega^2. \tag{10}$$

Due to this correlation the short noise is decreased at 30-70%. A similar correlation can be found in TES decreasing Johnson noise.

2.1.4 Ultimate noise performance of the CEB. General limit noise formula

This question about ultimate noise performance has arisen in relation to the highest requirements placed on NEP for future space missions. The question is, how realistic are these requirements on NEP=10^{-20} W/ Hz$^{1/2}$?

Ultimate performance of CEB and other bolometers has been analysed. Photon noise is not included in this analysis and should be added later as additional external noise. The NEP is determined by the shot noise due to the power load. The shot noise is treated in a general sense, including e-ph shot noise, due to the emission of phonons. Other sources of noise are neglected due to small values. For the level of P_0 =10 fW, this limit can be achieved using relatively low temperatures (~ 100 mK) and a small volume of the absorber ($\Lambda \leq 0.003$ μm^3) when we can neglect the electron-phonon noise component.

The general ultimate NEP formula for shot noise limitation has been derived (Kuzmin, 2004):

$$NEP_{shot} = (2P_0 E_{quant})^{1/2} \tag{11}$$

where P_0 – background power load, and E_{quant} –energy level of P_0 quantization:
$E_{quant} = k_B T_e$ –for normal metal absorber,
$E_{quant} = \Delta$ – for superconducting absorber.
Ultimate NEP can be estimated for different bolometers for relatively low power load $P_0 = 10\,fW$:

1. Type of bolometer	Characteristic parameter of absorber	2. Energy of quantization	3. Quantum efficiency $\hbar\omega / E_{quant}$	NEP$_{shot}$
CEB	T_e = 50 mK	$k_B T_e$ =4.3 µeV	950	$1*10^{-19}$ W/Hz$^{1/2}$
TES	T_C = 500 mK	Δ=73 µeV	56	$4*10^{-19}$ W/Hz$^{1/2}$
KID	T_C = 1.2 K (Al)	Δ=200 µeV	20	$7*10^{-19}$ W/Hz$^{1/2}$

Table 1.

The lowest NEP can be achieved for CEB with the lowest level of quantization. However, even these extreme parameters of P_0 and E_{quant} show that it is rather unrealistic to achieve NEP=10^{-20} W/ Hz$^{1/2}$ for P_0=10 fW.

Systems with linear on T thermal conductance

-    Spider-web TES with conductance through the legs,
-    CEB with cooling through SIN tunnel junctions (weak dependence on T: G ~T$^{1/2}$).
Limit shot noise is described by general formula (9) with numerical coefficient 2.

Systems with dominant e-ph thermal conductance (strong nonlinearity on T: G_{e-ph} ~T^4)

-    All bolometers on plane substrates with e-ph conductance.
-    Antenna-coupled TES on chip with Andreev mirrors.
-    NHEB with Andreev mirrors.

Due to strong nonlinearity of e-ph conductance, the limit shot noise is described by modified general formula (9) with five times increased coefficient 10.

As a common conclusion, if we leave the system for normal relaxation of energy through e-ph interaction, the shot noise is increased due to strong nonlinear dependence of electron-phonon thermal conductance on temperature in contrast to linear systems with weak dependence on temperature (or absence of it). This general formula (9) can be effectively used for estimation of the ultimate parameters of CEB and other bolometers for the given parameters of detector systems.

2.2 CEB in current-biased mode

2.2.1 Parallel/series array of CEBs with JFET readout

A novel concept of the parallel/series array of CEBs with superconductor-insulator-normal (SIN) tunnel junctions has been proposed for work under relatively high optical power load (10pW) and matching with the JFET readout (Kuzmin 2008a, 2008b) (Fig. 2). The current-biased CEBs are connected in series for dc and in parallel for HF signals. A signal is concentrated to the absorber through the capacitance of tunnel junctions and additional capacitance for coupling of superconducting islands. Due to the division of power between CEBs in the array and increasing responsivity, the noise matching could be effectively optimised and the photon noise equivalent power could be easily achieved at 300 mK with a room temperature JFET readout.

The CEB concept has been accepted as the main detector for the 350 GHz channel of BOOMERanG (Masi S. et al., 2006). The main requirement is to develop a CEB array with a JFET readout for the 90 channel system. The NEP of the CEB should be less than photon noise for optical power load of 10 pW and polarisation resolution should be better than 25 dB for observations of the CMB foreground polarisation.

The main innovation of the CEB array in comparison with a single CEB is the distribution of power between N series CEBs, summarising the increased response from the array. Effective distribution of power is achieved by a parallel connection of CEBs, which is coupled to the RF signal through additional capacitances (Fig. 4). The response is increased because the CEB is sensitive to the level of power and the power is decreased N times for the individual CEBs, with a proportional decrease of absorber overheating.

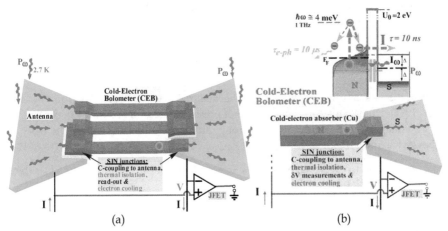

(a)   (b)

Fig. 4. (a) Schematic of a parallel/series array of CEBs with SIN tunnel junctions and JFET readout for the CMB polarization measurements. The current-biased CEBs are connected in series for dc and in parallel for HF signals through the capacitance of a SIN tunnel junction and additional capacitances; (b) The SIN tunnel junction is used also for electron cooling and for reading out the signal with a JFET.

The main purpose of this concept is to match the total dynamic resistance of the array to the noise impedance of a JFET (~0.6 MΩ). The power should be divided between the CEBs in the array to increase the responsivity due to lower overheating and moderate electron cooling. The high noise impedance of a JFET amplifier is one of the reasons why a low-ohmic TES (Irvin, 1995; Lee et al.,1996) could not be used for this application.

The operation of a CEB array can be analysed using the heat balance equation for a single CEB (1) taking into account power distribution between N bolometers:

$$\Sigma\Lambda\left(T_e^5 - T_{ph}^5\right) + P_{SIN}(V,T_e,T_{ph}) + C_v\frac{dT_e}{dt} = \frac{P_0 + \delta P(t)}{N} + 2\frac{V^2}{R_S} + I^2 R_A \qquad (12)$$

A bolometer is characterized by its responsivity, noise equivalent power and the time constant. In the current-biased mode, the responsivity, S_V, is described by the voltage response to an incoming power

$$S_V = \frac{\delta V_\omega}{\delta P_\omega} = \frac{\partial V / \partial T}{G_{e-ph} + 2G_{SIN} + i\omega C_\Lambda} \qquad (13)$$

Noise properties are characterized by the noise equivalent power (NEP), which is the sum of three contributions. For a series array of CEBs, the NEP is defined as follows:

$$NEP_{tot}^2 = N * NEP_{e-ph}^2 + N * NEP_{SIN}^2 + NEP_{JFET}^2. \tag{14}$$

Here NEP_{e-ph} is the noise associated with electron-phonon interaction (9); NEP_{SIN} is the noise of the SIN tunnel junctions. The SIN noise has three components: the shot noise $2eI/S^2_I$, the fluctuations of the heat flow through the tunnel junctions and the correlation between these two processes (Golwala, S. et al., 1997; Golubev & Kuzmin, 2001):

$$NEP_{SIN}^2 = \frac{\delta I_\omega^2}{(\frac{\partial I}{\partial V}S_V)^2} + 2\frac{<\delta P_\omega \delta I_\omega>}{\frac{\partial I}{\partial V}S_V} + \delta P_\omega^2. \tag{15}$$

This correlation is a form of the electrothermal feedback discussed earlier by Mather (Mather, 1982). Due to this correlation the shot noise is increased at 30-50% in contrast to the SCEB in voltage-biased mode where strong anti-correlation decreases the shot noise (Golubev & Kuzmin, 2002).

The last term is due to the voltage δV and current δI noise of the amplifier (JFET), which are expressed in $nV/Hz^{1/2}$ and $pA/Hz^{1/2}$:

$$NEP_{JFET}^2 = \frac{\delta V^2 + (\delta I * (2Rd + Ra) * N)^2}{S_V^2} \tag{16}$$

The strong dependence on N, decreasing this noise is included in the responsivity S_V, which is proportional to the N.

The proposed mode of CEB operation is a current-biased array with voltage readout by a JFET amplifier. The analysis of a single current-biased CEB with JFET readout has shown that there is no chance to get down to photon noise level for high power load due to decreased responsivity and JFET voltage noise (Kuzmin, 2008a, 2008b). The main reason is degradation of voltage responsivity under high optical power load due to overheating of the absorber.

The only chance to achieve a photon noise level is to use a series dc connection of bolometers. However, series HF connections of N bolometers would lead to real problems of junction size (capacitance should be increased proportionally to N) and overheating of islands. A special innovation has been proposed to combine these requirements: series connection for dc and parallel connection for RF. It could be realized by using additional capacitances for the HF coupling as shown in Figure 2(b). In this case, the input power is divided between bolometers, the electron temperature is decreased and the CEBs increase responsivity while the output signal is collected from all bolometers.

For power load of $P0 = 5$ pW per polarization, the photon noise could be estimated as

$$NEP_{phot} = \sqrt{2P_0 * hf} = 4.3 * 10^{-17} W / Hz^{1/2}. \tag{17}$$

We have simulated arrays of CEBs with different numbers of CEBs, from 1 to 26, to achieve a low NEP with JFET readout. The dependence of the noise components on the number of bolometers is shown in Figure 5. The total NEP decreases to a level less than photon noise for a number of CEBs larger than six (three for each probe). It is achieved mainly through the suppression of the JFET noise due to increased responsivity. Figure 5 demonstrates a strong linear increase of the responsivity proportional to N when the number of bolometers is increased. This dependence is well supported by linear asymptotic. The noise of the JFET is proportionally decreased, which is the main goal of this realization. Around the optimum point (N=22) the NEP_{JFET} is less than NEP_{SIN}, which is a manifestation of background-limited operation. The NEP_{SIN} increases proportionally to $\sqrt{N}$, but decreases due to decrease of the heat flow (and current) and an increase of the responsivity S. These two effects approximately compensate each other and NEP_{SIN} is not sensitive to the number of the bolometers. The most surprising result is that the NEP_{eph} (9) is not increased proportionally to the number of bolometers when the total volume of the absorber is increased proportionally to N. This is due to a compensation of this dependence by some decrease in T_e that is in the 6th power for NEPeph (9).

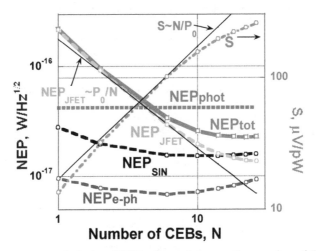

Fig. 5. NEP components and photon NEP in dependence on the number of CEBs in a series array. The parameters of CEB: I_{JFET}=5 fA/Hz$^{1/2}$, V_{JFET}=3 nV/Hz$^{1/2}$, R=1 kOhm, Λ=0.01μm^3. The responsivity S is shown for illustration of the effect of the CEB number. Thin lines show asymptotic for S and NEP_{JFET} (Kuzmin 2008a, 2008b).

The optimal number of CEBs in a series array is determined mainly by power load P_o and the volume of absorber Λ. The general rule of array design is the following: the number of bolometers, N, should be increased to split P_o between bolometers up to the point when $P_o/N = P_{ph}$, where $P_{ph} = T^5_{ph} \Sigma \Lambda$. The phonon power is determined by only one parameter, the volume of the absorber, Λ. There is no need to increase the number of bolometers more than this figure, because the optical power loading in each bolometer becomes less than the power from phonons. Responsivity is saturated beyond this level.

For a very small absorber volume, the optimal number of bolometers is determined by the interplay between amplifier noise and junction noise. The main rule here is to decrease the amplifier noise by increasing the number of CEBs to a level less than that of junction noise realising total noise less photon noise.

2.2.2 A two-dimensional array of CEBs with distributed dipole antennas and a JFET readout for polarization measurements

A novel concept of the two-dimensional array of CEBs with focal plane dipole antennas is proposed for sensitive polarization measurements (Kuzmin 2011c). The concept brings a unique combination of perfect polarization resolution due to a large uniform array of CEBs with optimal matching to JFET/CMOS amplifiers due to flexibility in dc connection. Better noise performance is achieved by distribution of power between the number of CEBs and increasing responsivity of bolometers. This arrangement should lead to substantial improvements in sensitivity and dynamic range. Reliability of the 2D array is considerably increased due to parallel/series connections of many CEBs. Polarization resolution should be improved due to uniform covering of a substrate by the 2D array over a large area and absence of beam squeezing to small lumped elements.

The fundamental sensitivity limit of the CEB array is below the photon noise NEP=$\sqrt{2hfP_o}$, which is referred to as the background-limited performance. Estimations of the CEB noise with JFET readout show an opportunity to realize background-limited performance with NEP less than photon noise NEP=5 10^{-17} W/Hz$^{1/2}$ at 350 GHz for an optical power load of 5 pW proposed for BOOMERanG.

The 2D array of CEBs with distributed dipole antennas (Fig. 6) is proposed for receiving one polarization components of the signal in multimode operation and effective matching to a JFET amplifier (Kuzmin 2008c). In this paper we analyse a realization of the CEB array for the 350 THz channel of BOOMERanG. The voltage response is measured by a JFET amplifier in a current-biased mode. The main purpose of the 2D configuration of the array with flexibility for dc connection is to achieve high responsivity (low noise) and to match the total dynamic resistance of the array to the noise impedance of a JFET (~0.6 MW).

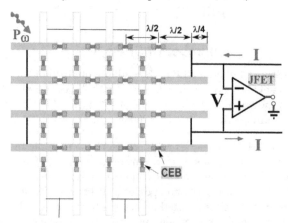

Fig. 6. Single polarization dipole antennas with a 2D array of CEBs (4x4) and a JFET readout. Each dipole antenna will be sensitive only to one polarization component of the RF signal. For illustration an array with a minimum number of 4x4 CEBs is shown with dc connection of 2x8 CEBs. Electrical isolation should be done between electrodes of the cross-dipole antennas.

Probability of failure is considerably decreased due to the parallel combination of several CEBs. For the analysed array we used four CEBs in parallel (W=4) and 16 rows of CEBs in series (N=16).

In contrast to Day et al. (2004), RF connection is made only for series connection in one line. Connection between lines is made only for dc bias by narrow wires.

Further improvement of performance is achieved by placing four bias resistors on chip just near the array of bolometers to decrease the noise of current-bias resistors and to protect scheme from interferences (Kuzmin et al., 1991; Kuzmin & Haviland, 1991). The RF matching is realized by the resistance of a normal absorber, which is independent of the tunnel junction parameters.

In what follows for analysis of a 2D CEB array we shall use the basic model of the CEB with strong electrothermal feedback due to electron cooling (Kuzmin & Golubev, 2002; Golubev & Kuzmin, 2001) and the concept of parallel/series array in current-biased mode (Kuzmin 2008a, 2008b). The operation of a CEB array can be analysed using the heat balance equation for a single CEB, taking into account power distribution between the NxW bolometers:

$$2P_{SIN}\left(V,T_e,T_{ph}\right)+\Sigma\Lambda\left(T_e^5-T_{ph}^5\right)+2\frac{V^2}{R_j}+I^2R_a=\left[P_0+\delta P(t)\right]\bigg/(W*N) \tag{18}$$

Here we assume that the SIN tunnel junctions are current-biased and the voltage is measured by a JFET amplifier. The responsivity S_V is described by the voltage response to an incoming power

$$S_V=\frac{\delta V_\omega}{\delta P_\omega}=\frac{2\partial V/\partial T}{G_{e-ph}+2G_{SIN}} \tag{19}$$

Noise properties are characterized by the noise equivalent power (NEP), which is the sum of three contributions:

$$NEP_{tot}^2=N*W*NEP_{e-ph}^2+N*W*NEP_{SIN}^2+NEP_{AMP}^2 \tag{20}$$

Here NEP_{e-ph} is the noise associated with electron-phonon interaction (9); NEP_{SIN} is the noise of the SIN tunnel junctions (15). The last term is due to the voltage dV and current dI noise of a JFET, which are expressed in $nV/Hz^{1/2}$ and $pA/Hz^{1/2}$:

$$NEP_{AMP}^2=\left(\delta V^2+(\delta I*(R_d+R_a)*N/W)^2\right)\bigg/(S_V/W)^2 \tag{21}$$

The estimations were made for the 350 GHz channel of BOOMERanG.

The optimal way for polarization measurements of two components in the same place is by orthogonal cross-combination of them connected by narrow dc wires for bias and readout connection as shown in Fig. 7. Optical matching could be organised by using Si substrate of resonance length $l/4$ with backshort behind the substrate (Fig. 7) or by using Si substrate with antireflection coating (Tran & Page, 2009). An additional Si lens or horn could be added for better concentration of incoming power in the 2D array of CEBs. The main criteria

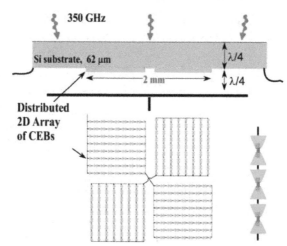

Fig. 7. Focal plane dipole antennas with a 2D array of CEBs consisting of two subarrays of 8x8 CEBs for each polarization. The RF matching is achieved by using Si substrate of resonance length $l/4$ with backshort behind the substrate.

are to realise high resolution of polarization at the level of 25 dB and to keep high efficiency of matching. The results of the simulation for the 2D array consisting of two sections of 8x8 CEBs for each polarization for different combination of dc connections are shown in Fig. 8. Fig. 8 shows that for all analysed dc combinations of width (W) and length (N) of the array we realize a total NEP of bolometer less than photon noise NEPphot. Fig. 8 shows that the optimal realization is for an array of 4x32. Degradation of performance for a wider array of 8x16 CEBs is related to a decrease of responsivity S=dV/dP.

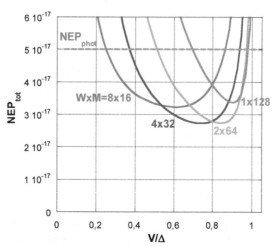

Fig. 8. NEP components of the 2D array of CEBs at 350 GHz, comprising two subarrays of 8x8 CEBs for each polarization (Fig. 4). Different combinations of CEBs (WxN) for dc connection are simulated for matching with the JFET readout. Power load = 5 pW, I_{JFET}=5 fA/Hz$^{1/2}$, V_{JFET}=3 nV/Hz$^{1/2}$, R=3 kOhm, L=0.01mm^3, T=300 mK.

Degradation of performance for more narrow arrays of 2x64 and 1x128 CEBs is related to an increase of total dynamic resistance of the array more than the noise impedance of a JFET amplifier. To compensate this mismatch, the optimal operation point is shifted to higher voltages with lower resistance. From these simulations the optimal array in the sense of noise and reliability is a 2D array with four CEBs connected in parallel and 32 CEBs connected in series.

The main progress in matching with the JFET readout is achieved by proper selection of the parallel and series combination of CEBs in 2D arrays, and selection of proper resistance of SIN junctions (Fig. 8). Results of simulation for the 2D array with optimal combination of 4x32 CEBs for various resistances of the SIN tunnel junction are shown in Fig. 9.

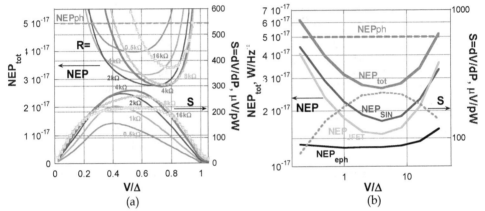

Fig. 9. (a) Voltage dependence of total NEP and responsivity S=dV/dP for the 2D array of CEBs, comprising two subarrays of 8x8 CEBs for one polarization with optimal combination of 4x32 CEBs for dc connection for different values of SIN junction resistance R; (b) Dependence of NEP components of the same 2D array on resistance of SIN tunnel junction. The parameters of CEBs are the same as in Fig. 8.

Optimal value of resistance for minimum NEP is 3 kOhm. Noise components in dependence on resistance are shown in Fig. 9b. The total noise is determined mainly by noise of SIN tunnel junction (15) and noise of amplifier (JFET) (21). The electron-phonon noise (9) is small due to the small volume of the absorber and low temperature independently on a rather large number of CEBs.

Finally, a novel concept of the 2D array of CEBs with distributed dipole Antennas has been proposed for polarization measurements with high resolution of polarization components. This concept provides more flexibility in matching to JFET readout and better noise performance due to the distribution of power between the number of junctions with decreasing power load and the increasing responsivity of each individual CEB. Polarization resolution should be improved due to the absence of beam squeezing to small lumped elements. The reliability of 2D array is considerably increased due to the parallel/series connections of many CEBs.

2.2.3 2D array of CEBs with distributed cross-dipole antennas for multimode measurements of both polarization components

A novel concept of the 2D array of CEBs with a cross-dipole antenna is proposed for ultra-sensitive multimode measurements of both polarization components of an RF signal (Kuzmin 2011a). This concept gives the opportunity of avoiding complicated combinations of two schemes to measure simultaneously both polarization components. The optimal concept of the CEB including a superconductor-insulator-normal (SIN) tunnel junction and an Andreev SN contact is used for better performance. This concept provides better matching with the JFET readout, suppresses charging noise related with Coulomb blockade due to the small area of tunnel junctions and decreases the volume of a normal absorber for further improvement of noise performance. The reliability of 2D array is considerably increased due to the parallel/series connections of many CEBs.

Estimations of the CEB noise with the JFET readout show an opportunity to realize NEP less than photon noise NEP=4 10^{-19} W/Hz$^{1/2}$ at 7 THz for an optical power load of 0.02 fW.

The CEB is a planar antenna-coupled superconducting detector that can easily be matched with any planar antenna. Promising direction is distributed focal plane antennas (Kuzmin, 2008c; Day et al., 2004). These antennas could help to avoid horns or Si lenses for matching with bolometers and could be used in a multimode regime for wide band spectrometers. To achieve RF matching to a distributed focal plane antenna, a series array of the CEB and JFET readouts have been analysed (Kuzmin, 2008c). However, this configuration has several disadvantages for the spectrometer: the antenna measures only one polarization component, resistance is too high for matching with the JFET and the Coulomb blockade starts to become important at low temperatures due to small junction capacitance, and the probability of failure is increased proportionally to the number of bolometers in the series array. To avoid these problems, a novel concept of the 2D array of CEBs with a distributed

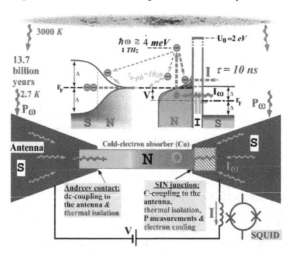

Fig. 10. Schematic of the optimal CEB with capacitive coupling to the antenna (Kuzmin, 2006). A normal nanoabsorber is coupled to the antenna through the capacitance of a SIN tunnel junction and through a SN Andreev contact. The SIN tunnel junction is used also for thermal isolation, electron cooling and for reading out the signal with a SQUID or a JFET.

cross-dipole antenna is proposed. The optimal CEBs with a SIN tunnel junction and Andreev contact (Kuzmin, 2006) is used to overcome the above-mentioned problems (Fig. 10). The system is proposed for SPICA and MILLIMETRON spectrometers, and other ultra-sensitive cosmology instruments.

Detection using CEB is obtained by allowing the incoming signal to pass from the antenna to the absorber through the capacitance of a SIN tunnel junction and through a SN Andreev contact. Using this optimal concept of CEB, we achieve several advantages: resistance is decreased for better matching with the JFET, the Coulomb blockade is suppressed due to the absence of the second junction creating SET transistor (Kuzmin & Likharev, 1987; Kuzmin & Safronov, 1988) and the effective volume of the absorber is decreased due to the proximity effect of the Andreev contact.

The main mode of CEB operation is a concept employing a 2D array of CEBs (Fig. 11) for effective matching to a JFET amplifier (Kuzmin, 2011). A distributed cross-dipole antenna (Fig. 11) is proposed for receiving both components of the signal in multimode operation. In this paper we analyse a realization of the CEB array for the 7 THz channel.

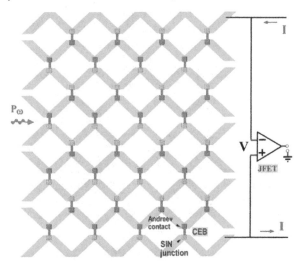

Fig. 11. Distributed cross-dipole antenna with a 2D array of CEBs and a JFET readout. This double polarization antenna is sensitive to both components of the RF signal.

The voltage response is measured by a JFET amplifier in a current-biased mode. The main purpose of the 2D array for readout is to match the total dynamic resistance of the array to the noise impedance of a JFET (~0.6 MΩ).

Probability of failure is considerably decreased due to the parallel combination of several CEBs. For the analysed array we used four CEBs in parallel (W=4) and eight rows of CEBs in series (N=8).

Further improvement of performance is achieved by placing four bias resistors on chip for decreasing noise of bias resistors and interferences (Kuzmin, 1991). RF matching is realised by the resistance of a normal absorber.

In the following analysis of a 2D CEB array we shall use the basic model of the CEB with strong electrothermal feedback due to electron cooling (Kuzmin & Golubev, 2002; Golubev & Kuzmin, 2001) and the concept of a parallel/series array in current-biased mode (Kuzmin 2008a, 2008b) similar to section 2.2.2. The operation of a CEB array can be analysed using the heat balance equation for a single CEB, taking into account power distribution between the NxW bolometers.

The estimations were made for the 7 THz channel of SPICA. Results of the simulation for the 2D array with the JFET readout are shown in Fig. 12. The Fig. 12 shows considerable improvement of noise properties for transition from the CEB with a double SIN junction to the CEB with a single SIN junction and an SN Andreev contact.

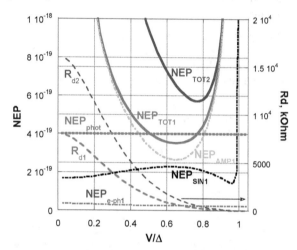

Fig. 12. NEP components for the CEB with a single SIN junction and an SN Andreev contact ("1") and for the CEB with a double SIN junction ("2"). Dynamic resistance, Rd, is shown for both cases referred to on the right axis. Parameters: I_{amp}=5 fA/Hz$^{1/2}$, V_{amp}=3 nV/Hz$^{1/2}$ (JFET), R=4 kOhm, Volume Λ=0.002um^3, T=70 mK, Tc =400 mK. The 2D array consists of four CEBs in parallel (W=4) and eight CEBs in series (N=8).

Improvement of NEP for single SIN junction is achieved due to the double decrease of dynamic resistance, Rd, and proportional decrease of the amplifier noise determined by a product of the current noise of the amplifier and the Rd (20). The total noise is determined mainly by the noise of the JFET amplifier (21) and the noise of the SIN tunnel junction (15).

The electron-phonon noise (9) is small due to the small volume of the absorber and low temperature. As we can see from Fig. 12, noise performance for the optical power load of 0.02 fW fits requirements with NEPtot<NEPphot for R=4 kOhm.

The simulations show that better NEP can be obtained with a decreased gap of Al to the level of 400 mK. This suppression of the gap can be obtained with additional evaporation of any normal metal (Cu, Ti,...) on the top of the counter electrode.

The results of the simulation for the 2D array for different resistances of the SIN tunnel junction are shown in Fig. 13. As we can see from Fig. 13, noise performance for the optical power load of 0.02 fW fits requirements of SAFARI with NEPtot<NEPphot for R>4 kOhm.

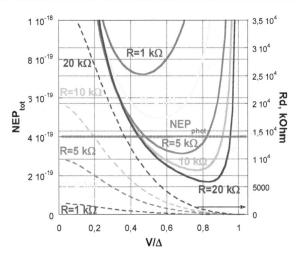

Fig. 13. NEP components of the 2D array of CEBs with the JFET readout at 7 THz with power load of 0.02 fW for I_{JFET}=5 fA/Hz$^{1/2}$, V_{JFET}=3 nV/Hz$^{1/2}$, Λ=0.002μm^3, T=70 mK, Tc =400 mK for different values of normal resistance R.

Improvement of NEP for larger R is achieved due to the increase of responsivity Sv=dV/dP proportional to R. For the left slope of NEP curves, this improvement is compensated by the increase of amplifier noise due to the product of the current noise of the amplifier and dynamic resistance of the junction. For the right slope of NEP curves, we do not have this limitation due to the noise of the amplifier and improvement is due to a slight decrease of overheating related with dissipation of power in a leakage resistance due to electron cooling.

The main progress in matching with the JFET readout is achieved by proper selection of the parallel and series combination of CEBs and selection of proper resistance of SIN junctions (Fig. 13). Some progress in NEP is realised due to replacing two SIN junctions by one SIN and one SN Andreev contact (Fig. 12). Some progress in better performance is achieved also due to proper suppression of Delta of the top superconducting electrode to the level of 400 mK (instead of 1.2 K for clean Al). The internal overheating of CEBs by applied voltage is decreased and we can realise arrays of any size (even with N>100).

2.3 CEB in voltage-biased mode

2.3.1 SCEB with SIS' tunnel junctions and Josephson junction and a SQUID readout

A novel concept of a superconducting cold-electron bolometer (SCEB) with a superconductor-insulator-weak superconductor (SIS`) tunnel junction and Josephson junction has been proposed (Fig. 14) (Kuzmin, 2008d). The main innovation of this concept is utilising the Josephson junction for dc and HF contacts, and for thermal isolation. The SIS` junction is used also for electron cooling and dc readout of the signal. The SIS` junction is designed in loop geometry for suppression of the critical current by a weak magnetic field. The key to this concept is that the critical current of the Josephson junction is not suppressed by this weak magnetic field and can be used for dc contact. Due to this innovation, a robust two layer technology can be used for fabrication of reliable structures. A direct connection of SCEBs to a four-probe antenna has been proposed for effective RF coupling.

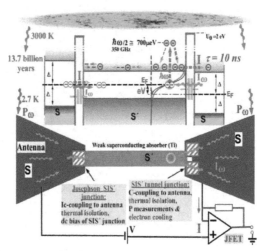

Fig. 14. Schematic of a SCEB with SIS` and Josephson tunnel junctions and a JFET readout. The SIS´ junction is used for capacitive coupling to the antenna, thermal isolation, electron cooling and dc readout by a JFET. The Josephson junction is used for dc and HF contacts, and for thermal isolation.

The main innovation of the SCEB in comparison with the CEB with double SIN junction (Kuzmin, 2000, 2001, 2004; Kuzmin & Golubev, 2002) and CEB with SIN and Andreev contact (Kuzmin, 2006) is effective utilization of the Josephson junction for dc and HF contacts, and for thermal isolation. The SIS` junction (for HF coupling, thermal isolation, electron cooling and dc readout) is proposed in loop geometry for suppression of the critical current by a weak magnetic field. Strikingly, the critical current of Josephson junction is not suppressed by this weak magnetic field. As a result of this innovation, a robust two layer technology can be used for fabrication of both SIS´ and Josephson tunnel junctions. In this paper we analyse realization of the SCEB for 350 GHz channel of BOOMERanG and other cosmology instruments.

Noise properties are characterized by the NEP

$$NEP^2_{total} = NEP^2_{e-ph} + NEP^2_{SIS'} + \frac{\delta I^2}{S_I^2}. \tag{22}$$

Here $NEP2e\text{-}ph$ is the noise due to electron-phonon interaction; $NEP2SIS´$ is the noise of an SIS´ junction and $\delta I^2/S^2_I$ is the noise of a JFET amplifier.

The noise of the SIS´ junction has three components: shot noise $2eI/S2I$, the heat flow noise and the correlation term between these two processes (Golwala, S. et al., 1997; Golubev & Kuzmin, 2001)

$$NEP^2_{SIS'} = \delta P_\omega^2 - 2\frac{\delta P_\omega \delta I_\omega}{S_I} + \frac{\delta I_\omega^2}{S_I^2} \tag{23}$$

This correlation is a form of the electrothermal feedback discussed earlier (Mather, 1982). For a superconductor absorber with concentration of electrons just near the gap, this anticorrelation is very strong and could lead, in first approximation, to almost 100% compensation of the shot noise:

$$\delta P_\omega^2 = 2\Delta P_0 , \ \delta I_\omega^2 = 2e^2 P_0 / \Delta , \ S_I = e/\Delta , \ NEP_{SIS'}^2 \cong 0 \tag{24}$$

For every chosen voltage we first solve the heat balance equation, find the electron temperature in the absorber, taking into account the effect of the electron cooling, and then determine current responsivity and *NEP*.

For optical power load of $P_0 = 5$ pW for each polarization of 350 GHz channel the photon noise is $NEP_{phot} = \sqrt{2P_0 * hf} = 4.3 * 10^{-17} W / Hz^{1/2}$.

Figure 15 shows a simulation of the different contributions to the total NEP of the detector for an optimised geometry of the bolometer. We see that for a range of bias voltage from 170 µV to 210 µV, the total NEP of the SCEB is well below the photon noise: $NEP_{tot} < NEP_{phot}$. The range of voltages less than 170 µV is not recommended for use because, due to negative slope of the IV-curve, the operation point would be unstable. In addition, the NEP_{tot} of the SCEB is dominated by shot/heat noise of the detector ($NEP_{SIS'}$) corresponding to a background limited mode of operation.

Figure 15b illustrates the effect of the noise reduction of SIS´tunnel junction. The figure shows all components of SIS´noise: NEP_{shot}, NEP_{heat} and correlation term $(dPdI)^{1/2}$. The final noise, $NEP_{SIS'}$, is clearly less than original noise components. The effect is stronger than for SIN junction noise (Golubev & Kuzmin, 2001) due to the well-defined level of quasiparticle energy just near the gap.

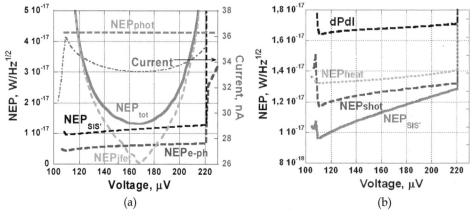

Fig. 15. (a) NEP components of the SCEB for I_{JFET}=5 fA/Hz$^{1/2}$, V_{JFET}=3 nV/Hz$^{1/2}$, R=0.2 kOhm, Λ=0.04um^3. The NEP_{tot} is less than NEP_{phot}; (b) Resulting SIS´ junction noise with strong cancellation of the NEP_{shot} and NEP_{heat} due to the anticorrelation term dPdI between them.

2.3.2 Parallel array of CEBs with distributed slot antennas and a SQUID readout

An innovative concept of the parallel array of CEBs with superconductor-insulator-normal (SIN) tunnel junctions has been proposed for distributed focal plane slot antennas (Kuzmin,

2008c). The parallel connection of CEBs with SIN tunnel junctions in voltage-biased mode is optimal for a slot antenna (Fig. 16). Some improvement of properties can be achieved by using optimal configuration of CEB with SIN junction and SN Andreev contact shown in Fig. 7 (Kuzmin, 2006). Any use of a double junction in voltage-biased mode (Kuzmin, 2001,2004; Kuzmin & Golubev, 2002) would lead to splitting power between two junctions and some degradation of responsivity.

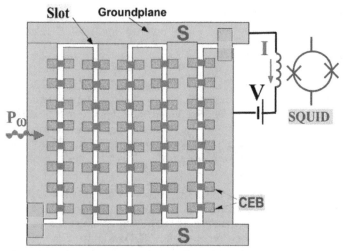

Fig. 16. A distributed single polarization slot antenna (Day, 2004) with a parallel array of CEBs (Kuzmin, 2008c) and a SQUID readout. This slot antenna will be sensitive only to horizontal components of the RF signal. The CEBs with SIN tunnel junctions and Al-Al SQUID could be fabricated on the same chip in one vacuum cycle.

An important feature of the design is that the volume of the normal metal is partly squeezed due to the proximity effect of the superconducting electrode of the Andreev contact. This squeezing further increases the efficiency of the electron cooling without degrading the HF coupling. Estimations of the CEB noise with the SQUID readout have shown an opportunity to realize background-limited performance for a typical power load of 5pW proposed for BOOMERanG.

The Al-Al SQUID could be fabricated on the same chip as CEB with SIN tunnel junctions in the same vacuum circle. Simultaneous fabrication of CEB and SQUID on-chip would create more reliable structures and avoid interferences due to wire interconnections of the systems.

We have analysed the concept of an optimal CEB for 70 GHz channel of B-Pol polarometer in the presence of the typical power load ($P_0 = 0.2$ pW).

Fig. 17 shows the results of a simulation of a CEB with a single SIN junction, with realistic parameters for the tunnel junction and absorber, and values of SQUID noise from 0.1 pA/Hz$^{1/2}$ to 0.8 pA/Hz$^{1/2}$. The level of NEPphot has been achieved for SQUID noise lower than 0.5 pA/Hz$^{1/2}$.

Optimal number of CEBs in series array. The optimal number (Fig. 18) is determined mainly by the power load Po and the volume of absorber Λ. The general rule of array design is the

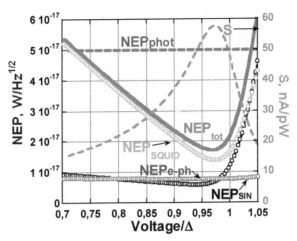

Fig. 17. Total NEP of the 40 CEB array with SIN tunnel junctions for the 350 GHz channel, with a SQUID noise current of 0.8 pA/Hz$^{1/2}$. Parameters:R=0.2 kOhm, S=1μm^2, Vol=0.005um^3, power load P_0 = 5 pW, T=300 mK. The NEPphot= 5*10^{-17} W/Hz$^{1/2}$ is shown by dashed line.per polarization component. For the 70GHz channel, NEPphot= 4.3*10^{-18} W/Hz$^{1/2}$.

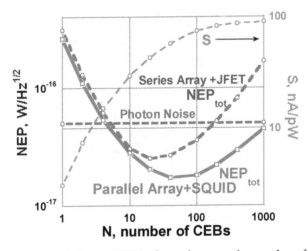

Fig. 18. NEP components and photon NEP in dependence on the number of CEBs in a voltage-biased parallel array with SQUID and in current-biased series array with JFET (section 2.2.2, M=1). The parameters of CEBs are the same as in Fig. 17. The responsivity S is shown for a parallel array for illustration of the effect of the CEB number.

following: the number of bolometers, N, should be increased to split Po between bolometers up to the point when $P_0/N= P_{ph}$, where $P_{ph} =T^5_{ph}$ $\Sigma\Lambda$. The phonon power is determined by only one parameter, the volume of the absorber, Λ. There is no need to increase the number of bolometers more than this figure because the optical power loading in each bolometer becomes less than the power from phonons. Responsivity is saturated after this level.

3. Experimental realization of the Cold-Electron Bolometer (CEB)

3.1 Technology of fabrication

The first CEB samples were fabricated using the Dolan shadow evaporation technique with a suspended mask made by e-beam lithography (Dolan, 1977). The technique is widespread for fabrication of sub-micron junctions for a wide range of applications. The technique utilizes a suspended mask fabricated using two-layer resist and shadow angle evaporation of two layers of metal. In the first CEB samples the aluminium electrode was thermally evaporated at an angle of 55° relative to the surface normal up to a thickness of about 60 nm (Kuzmin et al., 2002, Tarasov et al., 2002; Agulo et al., 2005). The tunnel barrier was formed by oxidizing the electrode for two minutes at a pressure of 5×10^{-2} mbar. The normal metal absorber was created by evaporating 30 nm of chromium and then 30 nm of copper at an angle of 0°. Cr was used to improve the impedance matching of the antennas to the normal metal and also for better adhesion of Cu to the substrate.

The first problem of this technology was related to evaporation of the normal metal as a top layer thicker than the base layer of Al. It led to rather high thickness of the absorber that degraded the performance of the bolometer. We tried an alternative technique, a so-called direct write technology with first depositing the normal metal absorber, its etching and oxidation and deposition of the superconducting Al layer after the second e-beam lithography (Otto et al., 2010). The disadvantage of this method is degradation of tunnel barrier properties during the second lithography process. To avoid this problem we modified the shadow evaporation technique by first depositing a bilayer of Cr/Al normal metal, oxidizing it, and then depositing the second superconducting Al layer (Tarasov et al., 2009).

The second serious problem was related to the strong limitation for an area of tunnel junctions in the Dolan shadow evaporation technique by submicron size that created serious problems with the capacitive coupling of the CEB and strength of electron cooling. The breakthrough in this problem occurred after the invention of the **Advanced Shadow Evaporation Technique (ASHET)** for large area tunnel junctions (Kuzmin, 2011b). This technique is dual to Dolan technique and gives the opportunity for fabrication of large area tunnel junctions (with area $\gg 1$ μm^2, really unlimited) in combination with nanoscale size absorbers.

3.2 Electron cooling experiments

The works on electron cooling are devoted to development of a CEB with capacitive coupling by SIN tunnel junctions to the antenna for sensitive detection in the THz region (Kuzmin et al., 2004; Agulo, I., et al., 2005). We used 4-junction geometry with Al-AlOx-Cr/Cu tunnel junctions and Au traps (Fig. 19).

The maximum decrease in electron temperature of about 200 mK has been observed at bath temperatures of 300-350 mK. Effective electron cooling was realized due to the improved geometry of the cooling tunnel junctions (quadrant shape of the superconducting electrode) and effective Au traps just near the junctions (≈ 0.5 μm) to decrease reabsorption of quasiparticles after removing them from normal metal.

Figure 20 shows the electron cooling curves and the theoretical fit for different temperatures for the structure with normal metal traps. In Fig. 20 we used for experimental curves both the experimental calibration curve of measured $V(T_{ph})$ and theoretical estimation of $V(T_e)$ shown in the same figure. The best coincidence of theoretical fit was obtained using the T^6-

temperature dependence. The experimental results show very good agreement with the theory for phonon temperatures in the range from 300 mK to 400 mK (Fig. 20).

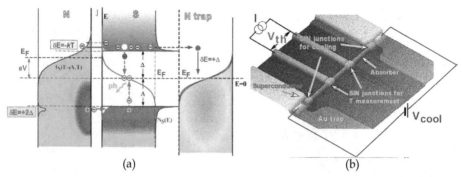

(a)   (b)

Fig. 19. (a) Energy diagram illustrates the principle of the electron cooling and the problem of reabsorption of the phonons after recombination of the quasiparticles. The normal metal trap is introduced to avoid this reabsorption. (b) The AFM image of the cooling structure was made using the shadow evaporation technique. The Au trap was evaporated prior to the evaporation of the Al-AlOx-Cr/Cu tunnel junctions.

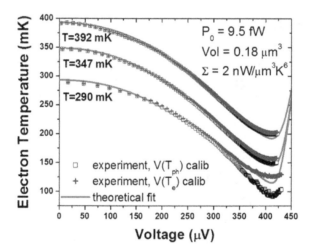

Fig. 20. Electron temperature measured for phonon temperatures of 392 mK, 347 mK and 290 mK. Theoretical fit for background power of 9.5 fW, absorber volume 0.18 μm^3.

3.3 DC qualification of the CEB

The electrical NEP measurements are realized in a special test structure with two junctions for electron heating and two junctions for temperature sensing (Agulo et al., 2005). In this case we can apply some small dc or ac heating power and measure the voltage response to such power dV/dP. The bolometer responsivity and the NEP were measured by applying a modulated heating current through the absorber. The frequency of modulation varied from 35 Hz to 2 kHz. The best responsivity of 15×10^9 V/W was obtained at 35 Hz. NEP of better than 2×10^{-18} W/Hz$^{1/2}$

was measured for modulation frequencies above 100 Hz. The bolometer time constant below 5 μs was estimated using the experimental device parameters (Agulo et al., 2005).

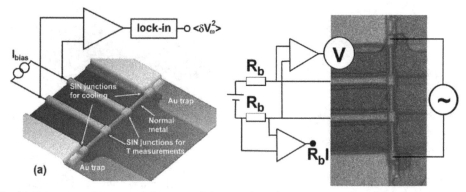

Fig. 21. Schematic diagram of the noise (left image) and responsivity (right image) measurements. The responsivity was measured by applying a heating power through the outer junctions with dc current and with modulated current, and then measuring the voltage response. The total noise was measured using a lock-in amplifier.

3.4 Quasioptical measurements

The responsivity of the parallel/series array of CEBs with a cross-slot antenna was measured using a blackbody radiation source equipped with thermometer and resistive heater (Tarasov et al., 2009; Tarasov et al., 2011). The blackbody radiation source (Fig. 23) was placed in front of the CEB attached to an extended hemisphere sapphire lens with antireflection coating. The source was mounted on a 3 K temperature stage of He3 cryostat. The temperature of blackbody can be varied in the range of 3-20 K.

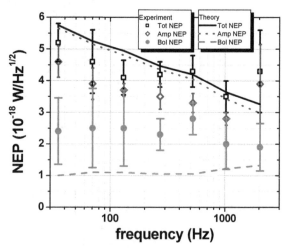

Fig. 22. Noise equivalent power spectrum from 35 Hz to 2 kHz. The bolometer NEP is at the level of $2*10^{-18}$ W/Hz$^{1/2}$, which agrees fairly well with what theory predicts.

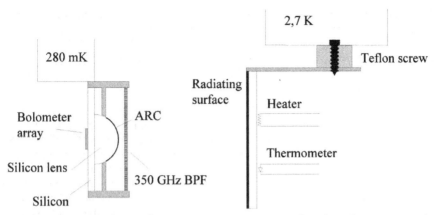

Fig. 23. Schematic view of optical response experiment. Here the silicon lens is covered with antireflection coating (ARC), incoming radiation pass through a band-pass filter for 350 GHZ (BPF). Thermal radiation source is equipped with heater and thermometer.

We measured the response to the microwave radiation emitted by a cryogenic blackbody radiation sources at the 100 mK stage of dilution refrigerator. The radiation source was mounted on the 0.4 K stage; it consists of a Constantan foil with heater and thermometer. Using a backward wave oscillator spectrometer/reflectometer we measured the reflection of the foil R=0.70±0.05 at 345 GHz. This value is different from zero reflectivity of blackbody and the actual emissivity of such source is κ=0.30±0.05. The response to heating of the emitter is presented in Fig. 24. The measured voltage response to temperature variations of the emitter is

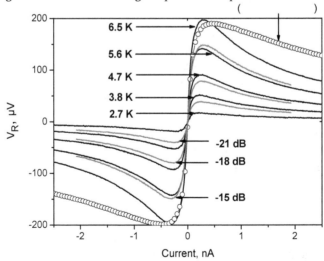

Fig. 24. Voltage response of the bolometer array to bath temperature changes 0.118-0.2 K (open circles), to variations of blackbody source temperature 2.7 K, 3.8 K, 4.7 K, 5.6 K, 6.5 K (solid lines) and also to 345 GHz radiation from BWO with additional attenuation of -21 dB, -18 dB and -15 dB (dashed lines).

25 µV/K. Taking into account the emissivity of foil and the root mean square voltage noise 0.38 µV in the frequency range 0-40 Hz, one can obtain the temperature sensitivity, which is 5 mK rms. If we assume that the noise spectral density of the amplifier $10\,nV/Hz^{1/2}$ dominates the total noise, we obtain a temperature sensitivity of $1.3*10^{-4}\,K/Hz^{1/2}$. We can also calculate the power emitted by the heated foil using the Planck formula for central frequency f_0=345 GHz and for bandwidth δf=100 GHz of a cross-slot antenna.

At a temperature of 3 K we get ΔP=$3*10^{-14}$ W, where h=$6.626*10^{-34}$ J*s is Planck's constant, k=$1.38*10^{-23}$ J/K is Boltzmann's constant, f is frequency and κ is emissivity of radiation source. The voltage response to incoming power is thus dV/dP=$8*10^{8}$ V/W. For the experimentally measured noise of $10\,nV/Hz^{1/2}$ this corresponds to an optical noise equivalent power NEP=$2*10^{-17}$ $W/Hz^{1/2}$. In Fig.14 we show that the responses of the detector to variations of the power from a thermal radiation source and from a BWO are very similar, while the response to changes of the physical temperature of the sample is clearly very different. This difference can be due to suppression of the energy gap due to thermal heating.

4. Conclusion

A novel concept of the capacitively-coupled CEB has been developed for high performance cosmology instruments. The CEB concept is based on a unique combination of *RF capacitive coupling* of an absorber to the antenna through capacitance of the SIN tunnel junctions and *direct electron cooling* of the absorber by the same SIN tunnel junction. The noise properties of this device are improved considerably by decreasing the electron temperature. Direct electron cooling leads to a considerable increase of the saturation power due to removing incoming power from the sensitive nanoabsorber.

Different concepts of antenna coupled CEBs from single bolometer to 2D arrays of CEBs help to match with various electrodynamic environments for polarization sensitive measurements and wideband FTS spectroscopy. Flexibility in the dc connection of CEBs in combination with electronic cooling give the opportunity to work with any optical power load from 0.02 fW to 10 pW realizing bolometer noise less than photon noise of the incoming signal. CEB can be matched with any readout from low-ohmic SQUIDs to high-ohmic JFET.

5. Acknowledgment

The author thanks Dmitry Golubev, Phillip Mauskopf, Peter Day, Paolo de Bernardis and Mikhail Tarasov for useful discussions. The work was supported by SNSB, VR, STINT and SI Swedish grants.

6. References

Agulo, I., Kuzmin, L, Fominsky, M., and Tarasov, M. (2004) "Effective Electron Microrefrigeration by Superconductor-Insulator-Normal Metal Tunnel Junctions", *Nanotechnology*, 15, 224-228.

Agulo, I., Kuzmin, L, and Tarasov, M. (2005) "Attowatt sensitivity of the Cold-Electron Bolometer". Proc. of the 16th Int. Symp. On Space Terahertz Technol., pp 147-152, Gothenburg, Sweden.

Bakker, J., Van Kempen, H. & Wyder, P. (1970). *Phys. Letts.* 31A, 290.

Chattopadhay, G., Rice, F., Miller, D., LeDuc, H.G. & Zmuidzinas, J. (1999). *IEEE Microwave Guide Wave Lett.*, 9,. 467.

Cho, A., Insights of the Decade: Precision Cosmology. *Science*, 330 p. 1615 (2010).

Day, P., LeDuc, H. G., Goldin, A., Dowell, C. D., & Zmuidzinas, J. (2004). Far-infrared/summillimeter imager-polarimeter using distributed antenna-coupled transition edge sensors. *SPIE* 5498, pp 857-865.

Dolan, G. (1977). Offset masks for lift-off photoprocessing. *Appl. Phys. Lett.* 31, 337.

Golubev, D. & Kuzmin, L. (2001). Nonequilibrium theory of the hot-electron bolometer with NIS tunnel junction. *Journal of Applied Physics*. Vol.89, pp. 6464-6472

Golwala, S., Johum, J. & Sadoulet, B. (1997). Proc. of the 7 International Workshop on Low Temperature Detectors, Munich, pp 64-65.

K. Irwin. (1995). An Application of Electrothermal Feedback for High-Resolution Cryogenic Particle-Detection. *Applied Physics Letters*, 66, (1995) 1998

Kuzmin, L. & Likharev, K. (1987). Direct Observation of the Correlated Discrete Single-Electron Tunneling, *Jpn. J. Appl. Phys.* 26 (suppl. 3), 1387 (1987).

Kuzmin, L. & Safronov, M. (1988). Observation of Single-Electron Coulomb Phenomena in Edge Tunnel Junctions, Pis'ma Zh. Eksp. Teor. Fiz. *JETP Lett.* 48,250 (1988).

Kuzmin, L., Nazarov, Yu.V., Haviland, D.B., Delsing, P. & Claeson, T. (1991). Coulomb Blockade and Incoherent Tunneling of Cooper Pair in Ultra-Small Junctions Affected by strong Quantum Fluctuations, *Phys. Rev. Lett.* Vol.67, 1161.

Kuzmin, L., & Haviland, D.B. (1991). Observation of the Bloch Oscillations in an Ultrasmall Josephson Junction, *Phys. Rev. Lett.* Vol.67, 2890.

Kuzmin, L., Pashkin, Y. A., Zorin, A. & Claeson, T. Linewidth of Bloch Oscillations in Small Josephson Junctions. *Physica B*, 203, 376, (1994).

Kuzmin, L. (1998) *Capacitively Coupled Hot Electron Microbolometer as Perspective IR and Sub-mm Wave Sensor*, Proc. of the 9th Int. Symposium on Space Terahertz Technology, Pasadena, pp 99-103.

Kuzmin, L., Devyatov, I. & Golubev, D. (1998). *Cold-electron bolometer with electronic microrefrigeration and the general noise analysis*. Proceeding of SPIE, v. 3465, San-Diego, pp. 193-199.

Kuzmin, L., Chouvaev. D., Tarasov, M., Sundquist, P., Willander, M., Claeson, T. (1999). On the concept of a normal metal hot-electron microbolometer for space applications. *IEEE Trans. Appl. Supercond.*, v. 9, N 2, pp. 3186-3189.

Kuzmin, L. (2000). On the Concept of a Hot-Electron Microbolometer with Capacitive Coupling to the Antenna, *Physica B*: Condensed Matter, 284-288, 2129.

Kuzmin, L. (2001) *Optimization of the hot-electron bolometer for space astronomy*, SNED Proc., pp. 145-154, Naples,,

Kuzmin, L., Golubev, D. (2002). On the concept of an optimal hot-electron bolometer with NIS tunnel junctions. *Physica C* 372-376, 378.

Kuzmin, L., Fominsky, M., Kalabukhov, A., Golubev D., & Tarasov M. (2002). *Capacitively Coupled Hot-Electron Nanobolometer with SIN Tunnel Junctions*, Proc. of SPIE, 4855, pp 217-227.

Kuzmin, L. (2003). Superconducting cold-electron bolometer with proximity traps. *Microelectronic Engineering*. 69, p. 309.

Kuzmin, L. (2004) *Ultimate Cold-Electron Bolometer with Strong Electrothermal Feedback*, SPIE Proc., 5498, p 349, Glasgow.

Kuzmin, L., Agulo I., Fominsky, M., Savin, A., Tarasov, M. (2004). Optimization of the electron cooling by SIN tunnel junctions, *Superconductor Science & Technology*, 17, pp. 400-405

Kuzmin, L. (2006). *Ultimate Cold-Electron Bolometer with SIN Tunnel Junction & Andreev Contact*. Proc. of the 17th Int. Symp. on Space Terahertz Technol., pp 183-186.

Kuzmin, L. (2008a). Array of Cold-Electron Bolometers with SIN Tunnel Junctions for Cosmology Experiments. *Journal of Physics: Conference Series (JPCS)*, 97, p. 012310.

Kuzmin, L. (2008b). A Parallel/Series Array of Cold-Electron Bolometers with SIN Tunnel Junctions for Cosmology Experiments, IEEE/CSC & European Superconductivity News Forum, No. 3, pp 1-9.

Kuzmin, L. (2008c). *Distributed Antenna-Coupled Cold-Electron Bolometers for Focal Plane Antenna*, Proc. ISSTT conference, pp 154-158.

Kuzmin, L. (2008d). A Superconducting Cold-Electron Bolometer with SIS´ and Josephson Tunnel Junctions, *Journal of Low Temperature Physics*, 151, pp. 292-297.

Kuzmin, L. (2011a). 2D Array of Cold-Electron Nanobolometers with Double Polarisation Cross-Dipole Antennas, *accepted to Nanoscale Research Letters*.

Kuzmin, L. (2011b). Advanced Shadow Evaporation Technique (ASHET) for Large Area Tunnel Junctions. US prepatent , Appl. No. 61/525242, filed August 19, 2011.

Kuzmin, L. (2011c). Two-Dimensional Array of Cold-Electron Bolometers for Ultrasensitive Polarization Measurements. *Radiophisika. Izvestiya VUZov*, т. LIV, N8-9, 607 (2011); *Radiophysics and Quantum Electronics* (2012).

Lee, A., Richards, P., Nam, S., Cabrera, B., Irwin, K. (1996). A superconducting bolometer with strong electrothermal feedback. *Applied Physics Letters*, 69, 1801.

Mather, J. C. (1982). Bolometer noise: nonequilibrium theory. *Appl. Opt.* 21,1125.

Masi, S. et al. (2006). Instrument, Method, Brightness and Polarization Maps from the 2003 flight of BOOMERanG, *Astronomy and Astrophysics*, 458,687-716, astro-ph/0507509

Nahum, M., Eiles, T.M. & Martinis, J.M. (1994). Electronic microrefrigerator based on a NIS tunnel junction. *Appl. Phys. Lett.*, 65, pp.3123-3125.

Otto, E., Tarasov, M., Kuzmin, L. (2007). Ti-TiO$_2$-Al normal-insulator-superconductor tunnel junctions fabricated in direct- write technology, *Supercond. Sci. Technol.* 20 865-869.

Tarasov, M., Fominsky, M., Kalabukhov, A., & Kuzmin, L. (2002). Experimental study of a normal-metal hot electron bolometer with capacitive coupling. *JETP Letters*, 76, pp. 507-510.

Tarasov, M., Kuzmin, L., Edelman, V., Kaurova, N., Fominsky, M. & Ermakov, A. (2010). Optical response of a cold-electron bolometer array. *JETP Letters*, 92, pp. 416-420.

Tarasov, M., Kuzmin, L. & Kaurova, N. (2009). Thin multilayer aluminum structures for superconducting devices. *Instruments and Experimental Techniques*, 52, pp. 877-881

Tarasov, M , Kuzmin, L.; Edelman, V.; Mahashabde S. and de Bernardis, P.(2011) Optical Response of a Cold-Electron Bolometer Array Integrated with a 345-GHz Cross-Slot Antenna, *IEEE Transaction on Applied Superconductivity*, 21, pp. 3635-3639.

Tran, H. & Page, L. (2009). Optical elements for a CMBPol mission. *Journal of Physics: Conference Series* 155 012007

4

Lens-Antenna Coupled Superconducting Hot-Electron Bolometers for Terahertz Heterodyne Detection and Imaging

Lei Liu

Department of Electrical Engineering, University of Notre Dame,
Notre Dame, IN,
USA

1. Introduction

Both astronomic observation and atmospheric remote sensing in the terahertz (THz) frequency range (0.1-10 THz) have driven the demand for highly-sensitive mixers and receivers [1-4]. Interstellar molecule spectrum information obtained through those THz receivers provides the basic clues to understand the formation of the universe. Also, the observation and monitoring of the earth's own atmosphere may allow strategies to be developed to address issues such as global warming or ozone depletion [5, 6]. Other terahertz radiation detection applications include plasma diagnostics [7], military radar and radiometry [8], chemical spectroscopy and analysis [9], bio-particle reorganizations [10, 11], security screening and terahertz imaging [12-14].

Two types of detection schemes are used in the terahertz regime, "direct" or incoherent detection (only magnitude information) [15, 16] and "heterodyne" or coherent detection (both magnitude and phase information). At terahertz frequencies, heterodyne detection (also called "mixing") typically exhibits higher sensitivity and greater dynamic range than direct detection schemes. Heterodyne receivers are frequency translators that convert a high-frequency ("RF") signal to a lower-frequency (called the intermediate frequency or "IF") band. The output IF signal is a replica of the high-frequency signal and preserves both the magnitude and phase information of the original signal.

Currently, THz heterodyne detectors (or receivers) based on superconducting niobium SIS mixers offer the highest sensitivity. However, these devices exhibit a gap frequency around 700 GHz at which the photon energy exceeds the binding energy of the Cooper pairs in the superconductor, directly converting the superconductor into its normal, lossy state. Above the gap frequency, superconducting hot-electron bolometer (HEB) mixers become much attractive since the operation of HEB does not depend on the tunneling of quasi-particles across the insulator.

Receivers employing Schottky diodes are routinely used in laboratories at room temperature. However, they require relatively large LO power ($\sim$1-10 mW), which is unpractical for space-borne applications and difficult to implement for receiver arrays. HEB receivers normally require very low local oscillator (LO) power ($\sim$100 nW), while providing higher

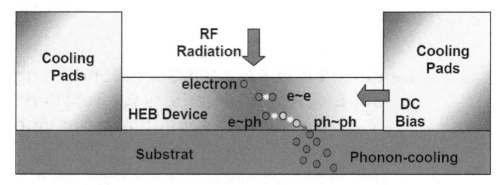

Fig. 1. The operation of HEB devices, the electron-electron cooling/scattering, electron-phonon cooling/scattering and phonon-phonon cooling/scattering are denoted as e-e, e-ph and ph-ph, respectively.

sensitivity than Schottky diode receivers, thus became an attractive research interest in recent years.

In this chapter, the basic theory/mechanism for HEB devices working as THz heterodyne mixing/detection elements will be first introduced, followed by a literature review on what have been done so far for quasi-optical (lens-coupled) HEB THz mixers. The state-of-the-art HEB mixer performance will be summarized. After that, the simulation, design, fabrication and characterization of 585 GHz hot-electron mixers based on annular slot antennas will be presented. On the basis of the single element mixer design, the realization and performance of one-dimensional and two-dimensional HEB mixer focal-plan arrays will be discussed.

2. Hot-electron bolometers and mixers

Hot-electron bolometer is a type of bolometric thermal detector that senses changes in temperature (T) through a change in resistance (R). Energy absorbed is distributed to an electron subsystem with a heat capacity C, and a thermal conductivity to a heat sink, G. Typically, the absorber (also the electron subsystem) is connected to a heat sink with a bath temperature T_b. The HEB voltage sensitivity S can be found by [17]:

$$S = I_{bias} \frac{dR}{dT} \frac{1}{G\sqrt{1 + \omega_{IF}^2 \tau^2}} \qquad (X.1)$$

where I_{bias} is biasing current, ω_{IF}, the IF frequency, and τ the thermal response time given by $\tau = C/G$. Note that in equation (x.1) the sensitivity is directly proportional to the change in resistance with temperature. Superconducting HEB's have a very sharp dR/dT slope around the critical temperature. Thus extremely sensitive receivers based on those devices can be achieved. Also, the heat capacity of electrons is much smaller than that of the lattice phonons. At low temperatures, coupling between the electrons and lattice is relatively weak, absorbed RF energy effectively heats only the electrons. Because coupling to the lattice phonons is weak, the lattice does not contribute much to the overall specific heat of the device, thus allowing faster cooling and broader bandwidth operation. This also results in a higher sensitivity, according to equation (x.1).

As seen in Fig. 1, heated electrons in the subsystem exchange absorbed energy (from RF power) through electron diffusion or phonon scattering. The characteristic electron-phonon scattering length L_{e-ph} (also called thermal heating length λ_{th}) is the mean free path before an inelastic electron-phonon scattering event takes place. If the bolometer length is larger than λ_{th}, the dominant cooling mechanism is phonon scattering, and the bolometer is called a "phonon-cooled" HEB. If the bolometer length is in the order of electron diffusion length L_{e-e}, but is less than λ_{th}, the bolometer is then called a "diffusion-cooled" HEB, and the dominant cooling mechanism is then diffusion of hot electrons to cooling pads connected to the device. As a result, the thermal response time for a diffusion-cooled HEB ($\tau_{e-e} \sim 0.1$ ns) is shorter than that of phonon-cooled HEB ($\tau_{ph} \sim 1$ ns), resulting in a broader 3-dB intermediate frequency (IF) bandwidth ($f_{IF,3-dB} = (2\pi t)^{-1}$) [17].

As discussed above, the small size (~ 100 nm) and fast response of superconducting HEBs allow them to be operated as heterodyne mixers (IF envelope detectors) rather than as simple power measuring devices. Fig. 2 shows how the HEB device operates as a mixer. The HEB absorbs power from the LO and RF signals. The device then warms, causing a portion of the HEB microbridge to become resistive (L_H). When operated in the optimal regime, the size and electron temperature of the resistive portion is very sensitive to the instantaneous power level. The HEB bridge cools by losing energy to the surrounding environment through various mechanisms as discussed above. The heating and cooling of the HEB follows the envelope of the LO and RF signals, which is at the IF (LO-RF or RF-LO) beat frequency. The resistance of the device, therefore, follows the beat signal between the LO and RF and thereby generates the IF when the HEB is current or voltage biased. This mixing mechanism allows HEBs to generate IF signals very effectively utilizing the material's superconductive/resistive transition at its critical temperature (T_c).

The canonical architecture for heterodyne mixing is shown in Fig. 3. The mixer usually comprises an electronic device (here will be HEB) with highly-nonlinear current-voltage (I-V) characteristics and its associated parasitic and matching networks. This is usually followed by a low-pass filter (LPF) and low noise amplifier (LNA) which, together, make up the IF chain. The frequency translation properties of a nonlinear element are readily illustrated by examining a square-law device with current-voltage relation: $I = AV^2$, with A being a constant dependent on the device. The RF input and LO can be represented as a superposition of sinusoidal functions at two slightly different frequencies, f_{LO} and f_{RF} as expressed in equation X.2.

$$V_{LO} = V_1 sin(2\pi f_{LO} t), V_{RF} = V_2 sin(2\pi f_{RF} t) \qquad (X.2)$$

When applied to the device, these signals will generate outputs at DC, $f_{RF} + f_{LO}$ and $f_{RF} f_{LO}$. Applying the square-law relation to the applied voltage yields the total current signal flowing through the mixer as

$$I = \alpha [V_{DC} + V_1 sin(2\pi f_{LO} t) + V_2 sin(2\pi f_{RF} t)]^2 \qquad (X.3)$$

This output signal, after filtering and amplification, will be proportional to V_1 and V_2, but is translated to a frequency of $f_{IF} = f_{RF} - f_{LO}$, as shown in equation X.4:

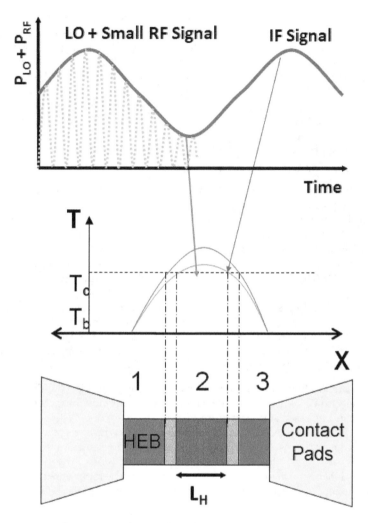

Fig. 2. The operation of a superconducting hot-electron bolometer device as a heterodyne mixer (figure reproduced from [1].

$$V_{OUT} = V_1 V_2 cos(2\pi f_{IF}t) \qquad (X.4)$$

Consequently, information contained in the signal at the RF frequency is down-converted to the IF band, which is more readily processed. This particular receiver scheme responds to both the RF signal and its image, f_{IM}, at $f_{IF} \pm f_{LO}$. Such a receiver is called a double sideband (DSB) mixer. In some cases, a band-pass filter (BPF) may be placed in front of the mixer, and reject the unwanted sideband. Thus the mixer is operated in single side band (SSB) mode and either the lower side band (LSB) or upper side band (USB) is chosen, depending on the overall application. Usually DSB mixers are undesirable because noise enters the image port

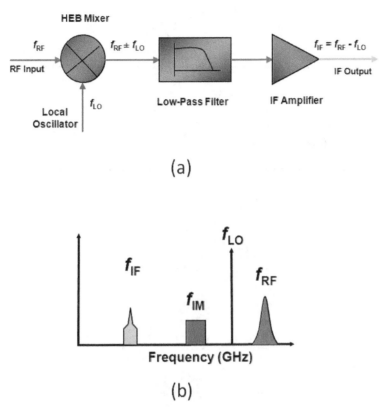

Fig. 3. Heterodyne mixing schematics: (a) Receiver diagram, (b) Signal spectrum.

and contributes to mixer noise. However, they are the most common type since it is difficult to implement image rejection filters at THz frequencies.

3. Lens-antenna coupled HEB THz mixer

3.1 Lens-coupled antenna configuration for THz detection and imaging

As the frequency increases into the submillimeter-wave and terahertz region, the detection of RF signals becomes challenging largely because of difficulties associated with extending classical microwave technologies and techniques to this frequency regime [18].   Loss introduced by metallic waveguides that are generally used at microwave frequencies increases with frequency (e.g. the surface resistivity of the waveguide is proportional to $\sqrt{\omega\mu/2\sigma}$). Also, the cost and tolerances associated with machining small structures are difficult to fulfill in fabricating waveguide structures with traditional milling techniques for use at terahertz frequencies [18]. Although microfabrication techniques using DRIE or SU-8 have been applied to the submillimeter-wave and terahertz waveguide structures [19], an alternative approach Ǔ quasi-optics is attractive, particularly as the wavelength of the signals approach the infrared region of the spectrum. Quasi-optical technology combines both the microwave and optical

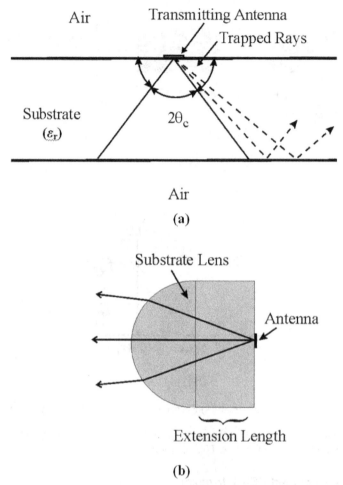

Fig. 4. (a) Transmission antenna on a dielectric substrate showing the generation of surface waves, reproduced from [20]. (b) Lens-antenna coupled THz detector and imaging configuration.

techniques, making it quite suitable and useful for applications in submillimeter-wave and terahertz region.

For submillimeter-wave and terahertz applications such as imaging arrays, that make use of quasi-optical technology, detectors or detector arrays are typically fabricated on a dielectric substrate such as GaAs, silicon, or quartz. The signal to be detected is first focused onto the substrate by an objective lens, and then illuminates the detectors from the dielectric side. In this way, the coupling structures (or antennas), should exhibit the maximum receiving directivity. However, antennas on dielectric substrates generate trapped surface waves, which unavoidably decrease the antenna efficiency and increase the cross-talk between adjacent antennas in an array [20]. This can be understood on the basis of simple geometric optics with a transmitting antenna shown in Fig. 4(a) [20]. The transmitting antenna lies at the upper air-dielectric interface, and radiates into the substrate. From the ray-tracing point of view, only the rays with a radiation angle of $\theta < \theta_c$ can go through the lower air-dielectric interface, where $\theta_c = arcsin(\epsilon_r)$ is the well-known critical angle. Other rays are completely reflected and trapped as the surface waves. The power that is converted to surface waves can be very large, dramatically reducing the antenna transmission/receiving efficiency. In addition, due to the existence of trapped surface waves, crosstalk between adjacent antennas in an imaging array can be significant increased, limiting the resolution that can be achieved by an imaging array.

The simplest method for solving the problem of surface wave excitation is to mount a dielectric lens to the detector/antenna substrate as shown in Fig. 4(b). If the lens and the substrate have the same dielectric constant, most of the incident rays are then nearly normal to the air-dielectric interface, eliminating total internal reflection at this interface. An imaging architecture, the so called "reverse-microscope" concept, is based on the above consideration and was first proposed by Rutledge and Muha in 1982 [13]. In this architecture, both an objective lens and a substrate lens are utilized for coupling incident radiation (THz) onto the antenna structure, resulting in an approach capable of diffraction-limited imaging. In our research, an extended hemispherical substrate silicon lens is utilized (see Fig. 4(b)), and the antenna coupled HEB mixers and mixer imaging arrays are fabricated based on this "reverse-microscope" configuration. This architecture will be further discussed later in this Chapter.

3.2 Lens-coupled THz antenna and mixer performance

As discussed above, at terahertz frequencies, RF circuits based on waveguide transmission media are more difficult to implement and quasi-optical techniques become an attractive alternative. To couple the incident RF power to the nonlinear HEB mixing element, planar antennas such as bow-tie [21], double-slots [22, 23], spiral [24], and log-periodic [25], mounted on lens-coupled dielectric substrates are frequently utilized as shown in Fig. 5. However, Bow-tie antennas have a number of drawbacks in the THz region since they are not compact for single imaging element design and exhibit antenna patterns with maximum off the antenna bore-sight. THz mixer designs using other antenna geometries also have not yet been suitable for high resolution imaging applications. Annular-slot antennas are one type of planar structure that can efficiently couple incident power to a device located at the feed point. They have very compact and symmetric geometries, making themselves good candidates for large scale imaging arrays. Moreover, this antenna structure can be easily scaled up to submillimeter-wave and terahertz region. In the following sections, we will focus on THz HEB mixers and focal-plane arrays based on annular-slot antennas.

State of the art HEB mixer performance compared to SIS and Schottky mixers are summarized in Fig. 6 [26]. Typically, receiver noise temperature in the order of 10 times of quantum limit has been achieved with HEB mixers. A theoretic analysis for the noise temperature of a superconducting HEB mixer is presented in [27]. Johnson noise and thermal fluctuation noise are the two main noise sources. For an optimum operation of a niobium device, the noise contribution from the above two mechanisms can be as low as 0.4 K from Johnson noise and 22 K from thermal fluctuation noise.

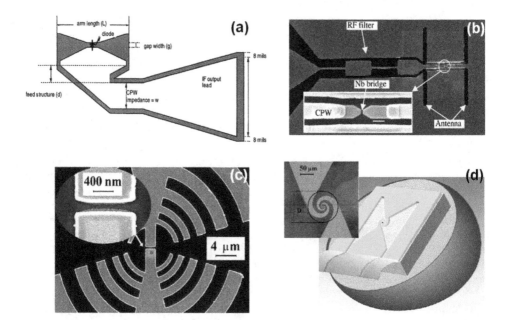

Fig. 5. Lens-coupled (a) bow-tie antenna [21], (b) double-slot antenna [22, 23], (c) log-periodic [25] and (d) spiral antenna [24] for THz HEB mixer design.

3.3 Superconducting hot-electron bolometer device fabrication

The HEB fabrication used in this research is based on the UVa EBL process developed by Bass et al [2]. The fabrication process includes several main steps: (1) base layer definition; (2), e-beam lithography (EBL) process to define the HEB length and width; (3) Reactive-Ion Etch (RIE) to remove unwanted metal, and (4), the passivation process.

The base layer definition begins with the deposition of a Nb/Au (10 nm/10 nm) bi-layer onto a square silicon wafer (high resistivity, 1.8 cm × 1.8 cm). This niobium base layer is used for defining the superconducting HEB micro-bridge, which is the most critical step in the mixer fabrication. To optimize the superconducting properties of this layer, a multi-target sputtering system is utilized and the deposition conditions (e.g. base pressure $\sim 10^{-8}$ Torr) are carefully controlled to achieve a low-stress niobium thin film. A film thickness of 10 nm is chosen to yield a sheet resistance of 35 Ω/square at the normal state and a critical temperature of $\sim$5.6 K [2]. A 10 nm gold over-layer is deposited with the same sputtering system to protect the superconducting niobium from oxidization during the subsequent fabrication processes.

After deposition of the Nb/Au bi-layer over the entire wafer, photoresist (AZ5206) is spun on top of the wafer as a sacrificial layer. The burn-off mask is first applied to eliminate the edge head, followed by the base layer lithography. Ti and Au (5 nm/200 nm) layers are then deposited onto this resist stencil using the e-beam evaporator system. Finally, a lift-off process is performed using NMP and propylene glycol (P-Glycol) heated to 110 °C to lift off (remove) the unneeded metals on the sacrificial layer, thus leaving the Ti/Au RF circuits and EBL alignment/focusing markers on the wafer. It is important to note that a gap in the Ti/Au

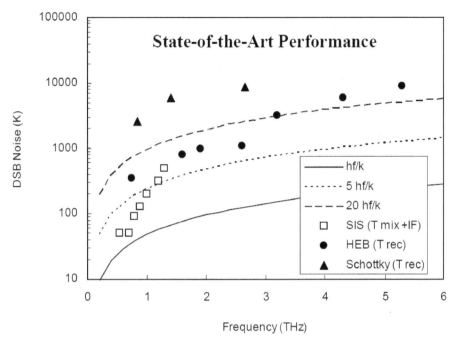

Fig. 6. State-of-the-art performance for SIS, HEB and Schottky mixers. This figure is reproduced from [26]

bilayer is left for the subsequent deposition of the HEB cooling pads and definition of the underlying 10 nm thick Nb HEB bridge.

After the base layer is defined, the mixing element – the HEB bridge – is fabricated using a two-step EBL process (see process flow in Fig. 7). In the first EBL step, a 200 nm thick bilayer PMMA (950/495) is spun on the base layer as the resist structure. The high molecular weight (950) PMMA on top is used for maximizing EBL resolution, while the low molecular weight (495) PMMA generates an undercut beneath the patterns after development, thus eliminating metal sidewalls during the following deposition process. This facilitates the lift-off process.

The HEB device cooling pads are then directly written by an electron-beam controlled by the Nano-Pattern Generation System (NPGS). During this "direct-writing" process, the EBL alignment markers on the base layer are applied to precisely "write" the patterns. Once again, prior to each HEB pattern writing, the focusing markers are utilized to adjust the SEM parameters to improve the device fabrication uniformity. To produce a HEB device length of around 200 nm, the cooling pads are designed using DesignCAD with a separation of 290 nm.

After developing the PMMA using a 1:3 mixture of MIBK and IPA, a trilayer of Ti/Au/Ti (10 nm/50 nm/10 nm) is evaporated. The first Ti layer serves as the adhesion layer and the second protects the Au during a subsequent RIE process. After a lift-off process using TCE heated to 70 °C, the HEB cooling pads are realized and the Nb microbridge length is defined. SEM pictures in Fig. 8(a) clearly show the designed and fabricated HEB cooling pads patterns. Each cooling pad overlaps both the Nb/Au HEB device layer film and the Ti/Au RF line and

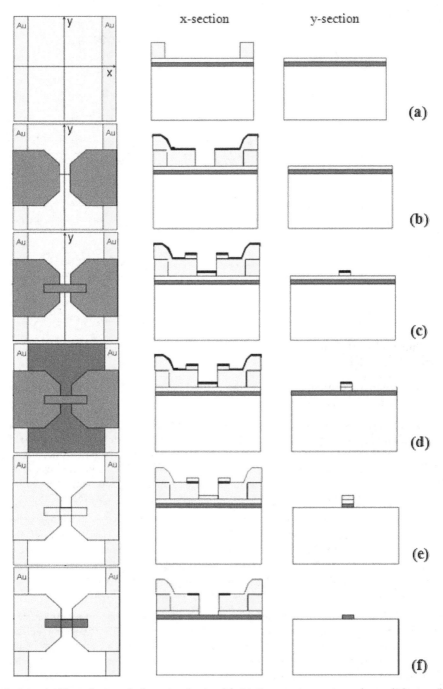

Fig. 7. A typical hot-electron bolometer device fabrication process using e-beam lithography.

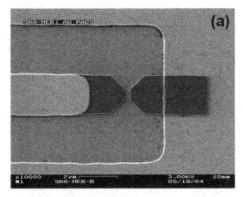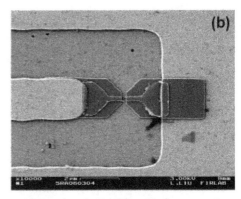

Fig. 8. Hot-electron bolometer device fabrication: (a)SEM picture showing the device cooling pads defining the HEB length, and (b)SEM picture showing the HEB bridge masks, defining the device width.

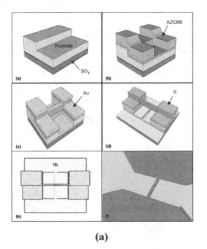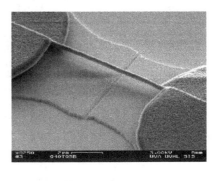

(a)   (b)

Fig. 9. Hot-electron bolometer device fabrication using the Ti-line technique without EBL process: (a) fabrication process, and (b) SEM picture of a suspended Ti-line (reproduced from [28]).

essentially serves to continue the Ti/Au contacts on each side in order to precisely define the length of the bridge.

During the second EBL step, the HEB bridge etch mask is first patterned by the NPGS with a single layer PMMA (950). The bridge width is designed to be 200 nm and 140 nm, for secheme-I and scheme-II structures, respectively. A Au/Nb (20 nm/20 nm) bilayer is deposited, and after lift-off, the bridge etch mask is left which also overlap part of the cooling pads as seen in Fig. 8(b). This Au/Nb bridge serves as a mask in the RIE processes, described below that defines and reveals the HEB Nb bridge width.

Reactive-ion etches (RIE) are therefore performed to define the width of the bridge to remove the unnecessary metal layers. First, an Ar-RIE is performed and after this etching, the open-field over-layer gold (10 nm) in the RF circuits is removed, exposing the underneath niobium layer (Fig. 7(d)). An SF_6-based RIE is then used to etch the exposed niobium layer both in the open field and on the top of the HEB bridge (Fig. 7(e)), followed by another Ar-RIE to remove the final layer of Au on top of the bridge, which leaves only a niobium bridge between the gold cooling pads (Fig. 7(f)). The conditions of the above etch steps are precisely controlled so that the etching rates are appropriate for operation. This is very important since over-etching typically significantly changes the device resistance and insufficient etching can results in total failure.

After the RIE process is finished, the HEB bridge is defined and the superconducting niobium is exposed to the ambient atmosphere. To prevent it from contamination and oxidization, a 300 nm germanium layer is e-beam deposited across the entire wafer. General photolithography is applied to have square photoresist patterns (20 μm $\times$ 20 μm) only in the HEB area. A subsequent SF_6-based RIE is performed to remove the exposed germanium, leaving square germanium covering the HEB devices. Since germanium is a semiconductor, carrier freeze-out and the low energy associated at sub-millimeter wave and terahertz frequencies prevents this material from shorting the RF circuits, while it provides a protective passivation layer for the superconducting niobium HEB devices.

Although the EBL-based HEB fabrication process is used in this research for prototype demonstration, it is strongly depends on the operator's skills and quite time-consuming. Thus the device uniformity is not perfect, and the process is not suitable for the fabrication of focal-plane arrays with large number of imaging elements. The so called "Suspended Sidewall Nano-Patterned Stencil" (SSNaPS or Ti-line) process as shown in Fig. 9 [28] is then strongly preferred. Using this method, the whole fabrication can be done without relying on those expensive e-beam or ion-beam facilities, allowing more research organizations to make and test HEB devices. In addition, quick fabrication of large quantities of HEB's on one wafer with perfect uniformity becomes readily realizable, making this SSNaPS process a good choice for making large focal plane arrays.

3.4 Single THz HEB mixer element based on annular-slot antennas

Annular-slot antennas are one type of planar structure that can efficiently couple incident power to a device located at the feed point. Thus these antennas have been studied in detail for a variety of microwave and millimeter-wave circuits including mixers [29] and frequency doublers [30]. Moreover, the annular slot antenna (ASA) provides a relatively compact geometry, which is attractive for high resolution imaging array applications [31], [32]. Fig. 10 illustrates the basic geometry of an annular slot with single feed-point and lying on a dielectric half-space with dielectric constant of ϵ_r. At its resonant frequency, the circumference of the annular slot is equal to nearly one wavelength, resulting in a sinusoidal electric-field distribution around the slot. As seen in Fig. 10(a), with incident field polarized along the y-direction, the electric-field in the slot has a maximum amplitude at $\varphi = 90\,°$ and $\varphi = 270°$, with nulls (or virtual shot-circuits) at $\varphi = 0°$ and $\varphi = 180°$. Thus the far-field radiation pattern has its E-plane along the y-axis, and H-plane along the x-axis. In summary, the annular slot features properties leading to its choice as a mixer coupling structure suitable for imaging array applications.

In this research, high-resistivity silicon is chosen as the mixer fabrication substrate because it has low loss ($\sim 2dB/cm$) and high dielectric constant ($\epsilon_r = 11.7$), which results in a high

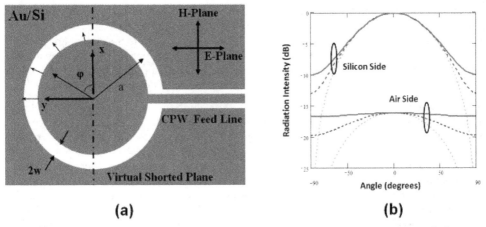

(a)   **(b)**

Fig. 10. (a) Annular-slot antenna on silicon substrate with circumference of one guided wavelength, for operation at 585 GHz. (b) Calculated radiation intensity of the 585 GHz annular-slot antenna.

directivity and efficiency for the receiving antenna [22]. The annular slot antenna is designed to operate at 585 GHz and has a radius, a, of 36 μm and a slot width, w, of 2.6 μm, as shown in Fig. 10(a). The annular slot antenna radiation pattern at 585 GHz in the silicon substrate and in the air side, for $\varphi = 0$, $\pi/4$, and $\pi/2$ planes are plotted in Fig. 10(b). As can be seen, the radiation intensity on the silicon side is much larger than that on the air side, resulting in a higher antenna directivity. The ratio of power radiated into the half-spaces is $\sqrt{\epsilon_{r-Si}}^3$: $\sqrt{\epsilon_{r-air}}^3$ ($\sim$ 40:1), according to [20].

An ADS Momentum simulation of the annular-slot antenna to determine its impedance has been performed, and the results show that the designed annular slot antenna has a bandwidth of approximately 100 GHz (giving a fractional bandwidth of 16%). At 585 GHz, the real and imaginary parts of the antenna input impedance are 100 Ω and approximately 0 Ω, respectively. Consequently, impedance matching to the devices is relatively straightforward because the superconducting HEB bridge is essentially a purely resistive device. A d-HEB with resistance as high as 100 Ω, however, requires nearly 3 squares of Nb thin film (10 nm thick with a sheet resistance of 35 Ω/square in the normal state). Hence, the resulting device length approaches that of the inelastic electron-phonon mean free path, L_{e-ph}. Because the resolution of the Nanometer Pattern Generation System (NPGS) employed at the University of Virginia is approximately 100 nm, the resulting device could well exceed the maximum length for diffusion cooling. As a result, a quarter-wavelength impedance transformer is employed to match the 100 Ω antenna impedance in the mixer design.

Another consideration in the receiver design using the lens-coupled antenna configuration is the Gaussian coupling efficiency and antenna system directivity. Analysis for double-slot antennas on extended hemispherical and elliptical silicon dielectric lens was carried out using a ray-tracing technique by Filipovic and Rebeiz et al. in 1993 [22]. Their results demonstrate that the directivity strongly depends on the extension length of the lens, L (see Fig. 4), and reaches a maximum at a particular value of L, while the Gaussian coupling efficiency maintains a relatively high value and then drops quickly as L increases. Thus, the design goal for an imaging array application is a high system directivity for maximum packing density in the focal plane, while maintaining an acceptable Gaussian coupling efficiency. To realize this design goal, the ray-tracing technique used by Filipovic is applied to the case of annular slot antennas on an extended hemispherical silicon lens. The hemispherical silicon lens for this work has a radius R = 4.5 mm. Radiation patterns of the lens-supported annular slot antenna are then calculated with a numerical computer code based on the ray-tracing technique.

The calculated E-plane patterns for various extension lengths are shown in Fig. 11. It is clearly seen that the main beam of the far-field pattern becomes narrower and then broader with increasing extension length. An extension length of 1600 μm is chosen for the highest antenna directivity for imaging array applications. In principle, once the antenna radiation patterns are obtained, the Gaussian coupling efficiency may be calculated with a double integration, as it is a parameter that measures the coupling efficiency between the far-filed patterns of the antenna and the incident Gaussian beam. However, this is very time-consuming due to the complex numerical integrations needed for the computer code. Since the results in [22] are given in terms of the parameters R/λ and L/λ, and these curves are generally applicable to other planer antenna designs with similar radiation patterns. With L = 1.6 mm, and L/λ = 0.36, the Gaussian coupling efficiency is estimated to be around 85%.

On the basis of the HEB mixer design and fabrication process described above, single element mixers have been fabricated and shown in Fig. 12(a). The one-square HEB microbridge is integrated at the end of the quarter-wave tranformer with a device length of 240 nm and a device width of 237 nm, resulting in a device resistance of nearly 35 Ω. In this circuit, high/low stepped-impedance low-pass filter is utilized to efficiently block the RF high-frequency signal. To characterize the superconducting properties of the fabricated HEB devices, R-T and I-V curves were taken, and shown in Fig. 12 in which (b) and (c) are the typical results [33]. As seen in Fig. 12(b), a sharp transition between the superconducting state and the normal state is observed at a critical temperature of T_c = 5.4 K with a transition width $\Delta T_c \sim 0.5K - 1.0K$. The typical critical current is measured to be approximately 120 μA, corresponding to a critical current density of $J_c \sim 5.1 \times 10^6 A/cm^2$. This value is comparable to the results reported by other groups [34]. The normal state resistance just above the superconducting transition is nearly 40 Ω, implying a thiner Nb film than intended, resulting in a slightly higher sheet resistance. Fig. 12(c) shows the I-V curves for a HEB device with slightly lower critical current. When the bath temperature increases from 4.2 K to 4.86 K, the critical current decreases from 70 μA to 55 μA, as anticipated.

To characterize the RF performance of the 585 GHz HEB mixers, a quasi-optical mixer mount was fabricated using oxygen-free copper for operation in a cryogenic vacuum system, as shown in Fig. 13. A tapered circular aperture in the front piece allows incoming radiation to fully illuminate the surface of the silicon lens. The thickness of this piece is designed so that the silicon lens is tightly held to the back piece, ensuring that there is no air gap between the lens and the substrate. A co-planar waveguide (CPW) transmission line was fabricated on a high-resistivity silicon wafter with a thickness of 1.1 mm, to accommodate the output IF signal (Fig. 13) [35]. The IF circuit substrate also serves as the extension length for the lens-coupled annular-slot antenna, and yields a total extension length of nearly 1.6 mm after mounting to the mixer chip. This extension length is close to the optimum value for maximum antenna directivity according to the ray-tracing results. The HEB mixer chip was mounted to the IF circuit using cryogenic epoxy, and the DC and RF connections were accomplished using gold bonding wires.

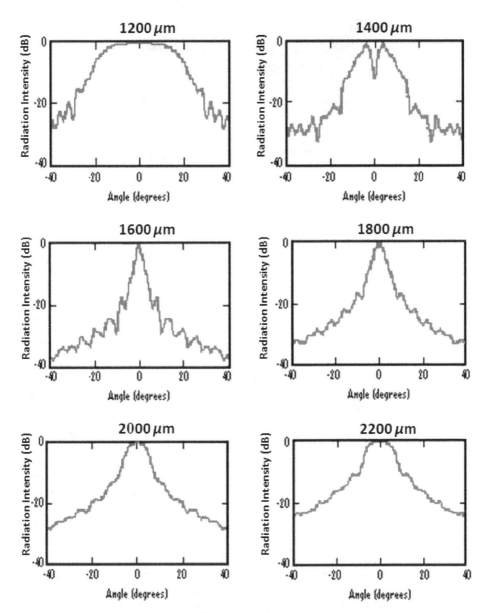

Fig. 11. (a) Annular-slot antenna on silicon substrate with circumference of one guided wavelength, for operation at 585 GHz. (b) Calculated radiation intensity of the 585 GHz annular-slot antenna.

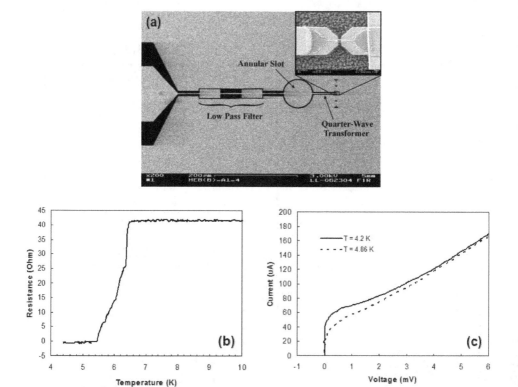

Fig. 12. (a) An annular-slot antenna coupled HEB mixer design utilizing a quarter-wave transformer for impedance matching. (b) R-T curve for a typical HEB device. (c) I-V curves of a HEB device measured at different temperatures.

Once the silver epoxy is cured, the HEB mixer is ready for testing. In this work, an HD-3(8) dewar system is utilized for the single element HEB mixer cryogenic tests and RF measurements. This dewar can be cooled to 4.2 K with a hold time of around 30 hours. Inside the dewar, the quasi-optical mount with the HEB mixer assembled is installed in front of a Teflon window and is biased through a bias-T. The IF signal is sent to an isolator (Durado 4ICB12-2) and low noise amplifier (LNA) before being fed to the external IF chain. Three temperature sensors are placed at the 4.2 K plate, LNA, and the quasi-optical mount and the temperatures are displayed outside of the dewar.

The RF measurement setup for characterizing the HEB mixers is shown in Fig. 14. A hot/cold load (300 K / 77 K) consisting of microwave absorber (Eccosorb) provides the blackbody RF radiation for a system Y-factor measurement. a VDI (Virginia Diodes, Inc.) 576 – 640 GHz FEM (Frequency Extension Module) is employed to provide an available LO power of 0.6 mW (or -2.22 dBm) near 585 GHz. The VDI FEM comprises three frequency doublers (D55v2, D100v3, D200) and a WR-1.5 frequency tripler. An Agilent E8247C (0-20 GHz) sweep oscillator and a power amplifier A246-2XW-31 (with a built-in doubler) are used for the signal input. To protect the VDI FEM, a WR-22 variable attenuator is inserted between the power amplifier output and the VDI FEM input. Once the FEM is appropriately biased, the output signal is

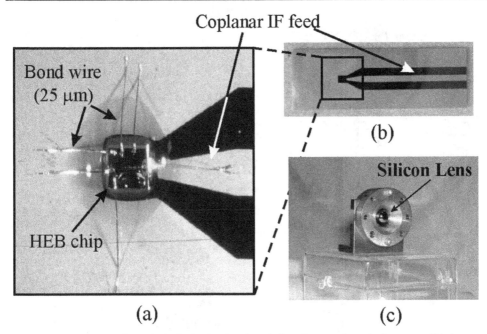

Fig. 13. Photographs of (a) the bolometer chip bonded in the quasi-optical system, (b) the coplanar IF feed for the antenna (on silicon), and (c) the cryostat lens mount. The chip is mouted directly to the IF feed circuit in (b) with a cryogenic epoxy (figure reproduced from [35].

obtained at a frequency of 48 times that of the input signal. For example, a 12.185 GHz signal with 16 dBm power level is used to produce a 585 GHz output from the FEM. According to the specifications provided by VDI, the output power of the FEM varies from 0.3 mW to 0.9 mW in the range of 576 to 640 GHz, which is more than adequate for pumping the HEB devices. Both the LO and RF are coupled into the cryogenic dewar through a set of lenses, beam splitter, and mirrors. Inside the dewar, the quasi-optical mixer mount is placed in front of the Teflon window and biased using a bias-T while the IF signal is output through an isolator and low noise amplifier (LNA) before being fed to the external IF chain.

With this measurement setup, the HEB pumped I-V curves at 585 GHz were initially measured by coupling only LO power into the dewar and to the device. As shown in Fig. 15, the un-pumped critical current for the HEB under test is 110 μA, close to the value measured previously. The critical current decreases with increasing LO power, as expected, and the HEB device is fully saturated (behaves like a pure resistance) at an LO power of P4, indicating that the entire Nb microbridge is pumped from the superconducting state into the normal state [35]. Note that the actual power received by the mixer is not measured directly, although it can (in principle) be estimated from the losses predicted in the system optics and RF circuitry. Nevertheless, The I-V curves measured demonstrate that adequate power is available for pumping the mixer into saturation. The measured E-plane radiation pattern of the lens-coupled annular-slot antenna shown in Fig. 15(b) demonstrates good agreement with calculation.

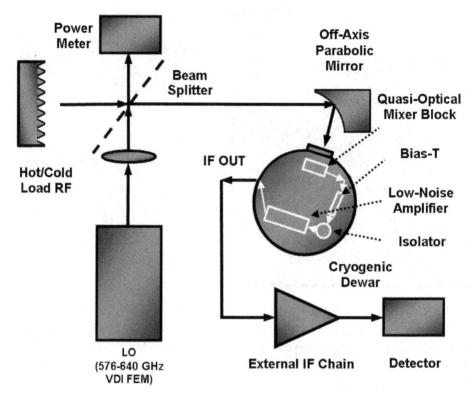

Fig. 14. RF and Y-factor measurement setup for characterizing the HEB mixer performance including gain and noise temperature.

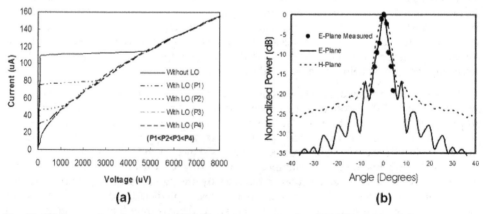

Fig. 15. (a) Pumped and un-pumped I-V curves measured at 585 GHz with a LO power provided by a VDI 576-640 GHz FEM. (b) Measured direct response vs. incident angles at 585 GHz compared to calculated antenna radiation E-Plane radiation pattern (H-plane pattern is also plotted using dotted line).

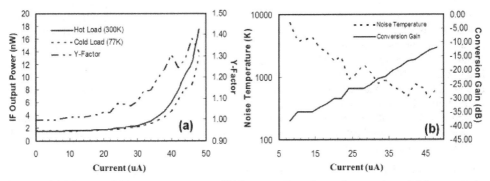

Fig. 16. (a) Measured IF output power and Y-factor as functions of current, and (b) corrected mixer conversion gain and SSB noise temperature vs. current.

Y-factor measurement was performed to characterize the HEB mixer gain and noise temperature. Figure 3.12 (a) shows the IF output power (at 1.8 GHz) together with the corresponding Y-factor at 4.3 K as functions of bias current in the case of a hot (300K) and cold (77K) load. It can be seen that the highest IF output power is measured at bias currents just before the device is fully saturated. Above this biasing point, the d-HEB becomes saturated and acts as a normal resistor. The highest Y-factor measured is approximately 1.20 dB, corresponding to an uncorrected DSB receiver noise temperature of $T_{rec,DSB} \sim 650K$ [35].

To calculate the mixer gain and noise temperature, the noise contribution from the RF optics and the IF chain must be removed from the measured data. Table 3.1 lists the gain and noise temperature contribution for the optical components. The total loss from the RF optics is estimated to be G_{rf} = -1.73 dB and the equivalent Planck noise temperature is thus T_{rf} = 114.6 K. The cryogenic LNA has a gain of G_{if} = 42 dB and a noise temperature of T_{if} = 3.0 K at 1.8 GHz. After correction, the effective noise level for the hot/cold (300K/77K) load at the input of the mixer is then 278.6K/128.8K ($T_{eff,hot} / T_{eff,cold}$). The mixer conversion gain is calculated by:

$$G_{mixer} = \frac{\Delta P_{out}}{\Delta P_{in} G_{if}} = \frac{\Delta P_{out}}{2kB(T_{eff,hot} - T_{eff,cold})G_{if}} \quad (X.5)$$

where ΔP_{in} and ΔP_{out} are input and output power change, k is Boltzmann constant, B is the IF bandwidth (B = 1.0 GHz for estimation since the external IF chain is not used for this measurement). The mixer DSB noise temperature can be found from,

$$T_{receiver} = T_{rf} + \frac{T_{mixer,DSB}}{G_{if}} + \frac{T_{if}}{G_{if}G_{mixer}} \quad (X.6)$$

according to equation 3.5, and the SSB mixer noise temperature is then $T_{mixer,SSB} = 2T_{mixer,DSB}$. The mixer gain and SSB noise temperature are shown in Figure 3.12 (b) as functions of bias current. Biased at 48 ţA and a bath temperature of 4.3 K, the mixer gain at 585 GHz is -11.9 dB and the SSB mixer noise temperature is $T_{mix,SSB} \sim 630K$ [35].

3.5 THz HEB mixer focal plane arrays for imaging applications

Highly-sensitive receivers employing superconducting hot-electron bolometers (HEB's) have been intensively studied and applied in millimeter-wave and far-infrared (FIR) imaging and remote sensing in recent years [1], [11]. However, in many applications, only one pixel of object information from receiver is insufficient. To effectively map the spatial distribution of radiation intensity, many pixels of imaging information are usually needed. Although mechanical scanning can be applied to a single element mixer to fulfill the above requirement, in some cases, this is not feasible due to the long observing time required to form a complete image [36], [37]. For example, in applications such as plasma diagnostics, information is often needed on a time scale of a microsecond, which is not usually possible with a single element detection system with mechanical scanning [18]. Imaging mixer arrays [13, 18] are the best approach for these applications since they can greatly reduce observing and processing time by recording imaging information in parallel.

Imaging arrays for submillimeter applications are a subject of continuing interest for both the astronomy and chemical spectroscopy communities. A number of researchers have put considerable effort into developing imaging arrays based on Schottky diodes or superconducting detectors [21, 39]. In 1982, D. B. Rutledge and Muha proposed a high-resolution imaging antenna array diagram with a "reverse-microscope" optical configuration [13]. On the bases of this diagram, a research group at the University of California at Davis is currently working on a 90 GHz Schottky diode mixer array with bow-tie antennas [21]. Bow-tie antennas, however, have a number of drawbacks for high resolution imaging applications in the THz region since they are not compact for single imaging element design and exhibit antenna patterns with maximum off the antenna bore-sight [20]. JPL has proposed a 1.6 THz 1-D array based on diagonal horns antennas [39]. However, 2-D arrays based on this scheme are difficult to realize. In recent years, another approach called the "fly's-eye concept" has been investigated at the University of Massachusetts at Amherst [38]. A 3-element HEB focal plane array based on this concept has also been reported recently demonstrating promising performance [38]. For this approach, each mixer element uses a separate imaging lens, which presents difficulties for design and fabrication, and hence the imaging resolution of this system will be limited.

Fig. 17 shows a diagram of HEB imaging array on an extended hemispherical silicon lens investigated in this work. Each of the imaging elements employs an annular-slot antenna as the coupling component into which a HEB device is integrated. Two difficulties have to be addressed to achieve a high-resolution imaging array. First, from the Shannon-Whittaker condition [40], the array element spacing should not exceed one wavelength at the frequency of interest for diffraction-limited imaging. Thus, very limited room is left for accommodating those circuits such as low-pass filters (LPF) and low-noise amplifiers (LNA). Second, while maintaining a small array element spacing is required to achieve high imaging resolution, mutual coupling between adjacent annular slot antennas and the cross-talk between the IF outputs become important and can limit increased resolution by simple scaling of the array geometry.

To achieve a diffraction-limited imagine with an HEB mixer array at submillimeter frequencies, the "reverse-microscope" optical configuration proposed by D. B. Rutledge and Muha [13] is utilized. Figure 18 shows a diagram of this configuration in which the annular slot antenna array substrate is attached to the back side of the imaging lens and an objective lens is placed in front of the imaging lens. The image signal is focused through both lenses onto the mixer array. The image of the object is then reconstructed by plotting the IF output signal from each element in the array. As discussed in previous sections, by utilizing the same material (silicon) for both the array substrate and the imaging lens in this configuration, the

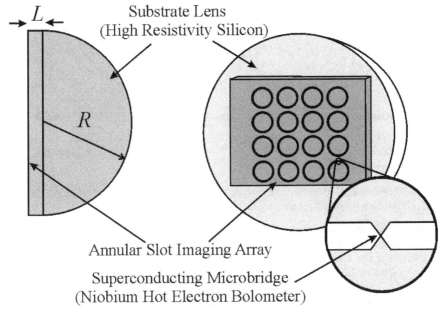

Fig. 17. Schematic of a HEB mixer imaging array using annular-slot antenna mounted on an extended hemispherical silicon lens with radius R (extension length L).

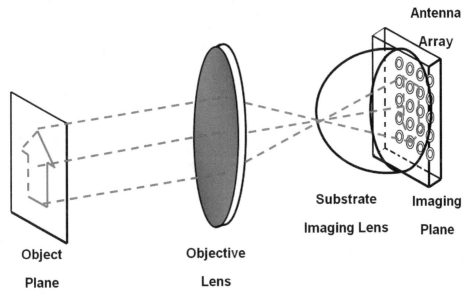

Fig. 18. Diagram of the reverse-microscope configuration proposed by D. B. Rutledge [13]. This configuration provides a capability of diffraction-limited imaging at THz frequencies.

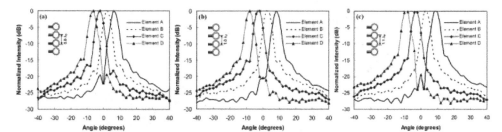

Fig. 19. Off-axis element radiation patterns for the lens-coupled one-D annular-slot antenna arrays calculated using ray-tracing technique. Patterns are calculated with element spacings of (a) 0.8 λ_d, (b) 1.0 λ_d, and (c) 1.1 λ_d.

trapped surface-wave is eliminated and the cross-talk between adjacent imaging elements is reduced, allowing a high imaging resolution to be achieved [13].

The single element HEB mixers employing annular slot antennas developed and fabricated in the earlier work in this research have shown promising performance. Here, the basic mixer design including the antenna, impedance matching circuits and low-pass filters are applied and expanded to realize a full imaging mixer focal-plane array. Due to the small element spacing ($\sim \lambda_d \approx 150\mu m$) for a diffraction-limited imaging, the mutual coupling between adjacent annular slot antennas in the mixer array is investigated to check whether it needs to be included in the design considerations. Both ADS Momentum simulations and EMF analysis [41, 42] were performed to study the mutual-impedances of the annular-slot antenna array with various element spacing, and a conclusion has been drawn that when the element spacing is larger than 0.8 λ_d (while smaller than 1.0 λ_d), the antenna mutual impedances and hene the cross talk between the adjacent elements are small enough that their effect can be safely neglected in the design of a diffraction-limited imaging system.

The element annular slot antenna (lens-coupled) off-axis radiation patterns for various spacings in prototype one-D (1×4) and two-D (2×2) arrays have been calculated using the ray-tracing technique. As shown in Fig. 19, the element antenna pattern in the one-D arrays has a 3-dB beam width of $\theta_{3-dB} \sim 4°$ with side-lobe levels less than -10 dB. The beam spacings between adjacent antennas are $\Delta\theta \sim 4.0°, 5.0°$, and 5.7° with crossover power level around -3.9 dB, -6.0 dB, and -7.0 dB for d = 0.8λ_d, 1.0λ_d, and 1.1λ_d, respectively. For imaging applications, a crossover level larger than 3 dB is generally used as a figure of merit to distinguish between two points and quantify the spatial resolution of an array. Fig. 20 shows the annular slot antenna off-axis radiation patterns in the 2 ×2 arrays. The element antenna pattern in the azimuth scan has the same properties as that in the one-D array. The element antenna pattern in the elevation scan has a 3-dB beam width of $\theta_{3-dB} \sim 3.5°$ with side-lobe levels less than -10 dB. The beam spacings between adjacent antennas are $\Delta\theta \sim 4.0°, 5.0°$, and 6.0° with crossover power level around -7.0 dB, -11.0 dB, and -12.5 dB for d = 0.8λ_d, 1.0λ_d, and 1.1λ_d. The antenna patterns in the elevation scan (or the E-plane pattern) have a narrower 3-dB beam width, compared to the patterns in the azimuth scan (or the H-plane pattern).

Shown in Fig. 21 are the fabrication and assembling results of the HEB imaging mixer arrays. For typical one-D arrays in Fig. 21 (a), quarter-wavelength impedance transformers were utilized. While twin-HEB devices were integrated in the two-D arrays due to limited spacing between adjacent antennas. The twin-HEB device has two HEB microbridges fabricated in

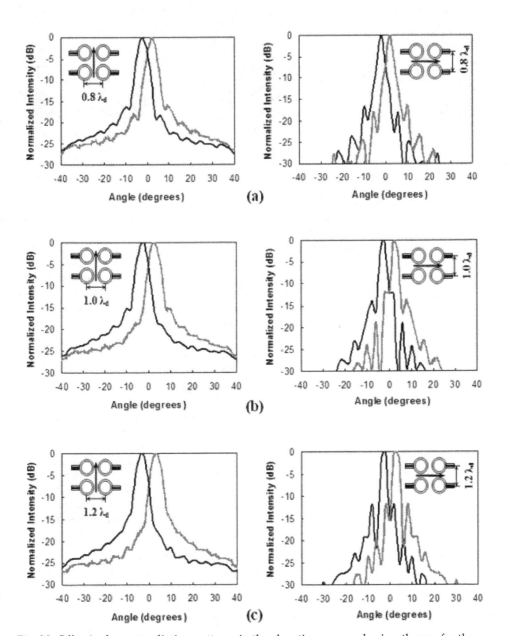

Fig. 20. Off-axis element radiation patterns in the elevation scan and azimuth scan for the lens-coupled two-D annular-slot antenna arrays calculated using ray-tracing technique. Patterns are calculated with element spacings of (a) 0.8 λ_d, (b) 1.0 λ_d, and (c) 1.2 λ_d.

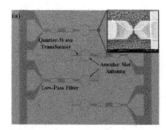 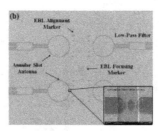 

Fig. 21. Lens-antenna coupled HEB mixer imaging array fabrication and assembling results: (a) SEM picture of an one-D HEB imaging array with quarter-wave impedance transformer, b) SEM picture of a two-D HEB imaging array using twin-HEB devices, and (c) the assembled mixer array mount, ready for cryogenic testing.

series (each with a normal-state resistance of 50 Ω) and connected by a gold cooling pad in bewteen, offering a device impedance of 100 Ω (matching to the antenna imbedding impedance). This approach eliminates the need for a lossy transmission line but requires a slightly more complex fabrication process. To measure the array performance, a quasi-optical mixer array block has been fabricated. As shown in Fig. 21 (c), the HEB mixer array chip was first bonded to a silicon substrate ($\sim$ 1.1 mm thick with four bended CPW transmission lines) using cryogenic epoxy, and an ultrasonice wire-bonding tool is used t oelectrically connect both the grounds and center conductors. Two other silicon substrates with CPW trsnsmission lines have been installed into the backside of the array block, and four subminiature A (SMA) connectors are utilized to output the IF signals.

A close-cycled cryocooler has been employed in the mixer array RF measurements. Inside the cryocooler as shown in Fig. 22, a 585 GHz mesh-filter ($\sim$40 GHz bandwidth) is placed on the 40K stage just in front of the Teflon window. The assembled mixer array block is installed with four outputs connected to the four bias-T's. The four-way Miteq power combiner, Durado cryogenic isolator and Miteq low-noise amplifier are anchored to the 4.2K stage with specially designed copper mounting pieces for good thermal contact. The RF components are connected using semi-rigid coaxial cables. An IF feedthrough is designed to couple the microwave signals out of the cryocooler. For better thermal insulation, the inside coaxial cable is coiled and the cable jacket is soldered to the copper heat sink, which is anchored to the 40K stage. After assembling all the inside components, the unused windows of the cover plate are covered with aluminum foils for further blocking the room-temperature radiation. During the HEB mixer array operation, only one of the four HEB devices is dc-biased at one time, and thus, each mixer can be measured separately.

Prior to RF testing, the DC charactersitcs of an one-D array with four mixer elements (A-D) were measured. A sharp transition from the superconducting state to the normal state is observed at around 5.5 K for elements A, B and C with $\Delta T_c \sim 0.5K$. The array element D became an open circuit at a temperature of $\sim$190K because the bond wire lost contact to the center conductor of the CPW transmission line due to the mechanical vibration from the close-cycled cryocooller system. The current-voltage (I-V) curves are then measured at a bath temperature of $\sim 3.9K$, as shown in Fig. 22 (b). The critical currents were measured to be 290 μA, 300 μA and 400 μA for element A, B, and C, respectively.

To measure the HEB mixer FPA imaging angular resolution, two adjacent elements (element B and C with B in the upper position) in an one dimensional array chip (see inset of Fig. 21(c), element spacing $d = 1.0 \ \lambda_d$) were DC biased and the responses were monitored. During the experiment, the VDI 585 GHz solid state source is employed and the distance between the silicon lens and the source is nearly 50 mm for maximum response. The responses for element B and C are normalized and plotted in Fig. 22(c) together with the off-axis antenna patterns (see Fig. 19) predicted using the ray-tracing technique. Again, reasonable agreement has been obtained. The discrepancies can be attributed to errors in source position measurement and the nonlinear relationship between the current response and the absorbed RF power. The vertical distance between the two maximum response position is 4.4 mm, resulting in an angular resolution of 5.04°, which is very close to the theoretical prediction (see Fig. 19).

By using an optical lens as an objective lens, the system imaging resolution can be further improved. Experiments have been performed with a distance of 133.1 mm between the silicon lens and the 585 GHz solid-state source. The optical lens is placed 49.3 mm from the cryocooler window (86.3 mm from the silicon lens). The response peaks from the element B and C are measured at source displacement of 69.35 mm and 72.10 mm, corresponding to an imaging resolution of ~2.75 mm. Compared to the results reported by the University of Massachusetts group (38) at 1.6 THz with a three-element HEB array based on the "fly's-eye" concept [38], the integrated HEB mixer arrays on the basis of the "reverse-microscope" configuration provide a capability to realize diffraction-limited, high-resolution focal-plane arrays in the THz frequency region.

Y-factor measurements have been performed to each of the two array mixer elements (element B and C) and the results show that DSB mixer noise temperatures of 1675 K and 3517 K, with mixer conversion gain of -14.73 dB and -17.74 dB, respectively, have been obtained for the two adjacent elements in the one dimensional focal-plane mixer array, which is comparable to the results reported in the literature. To the author's best knowledge, the HEB mixer focal plane array described in this paper is the first heterodyne FPA reported on the basis of the "reverse-microscope" architecture with the capability of diffraction-limited imaging. With this architecture, all the array antennas, HEB devices and IF circuits are integrated onto one silicon wafer, thus providing an alternative way for developing large FPA's in the THz region.

The measurement results in Section IV (B) are comparable to that reported in the literature [38]. However, the measured noise temperatures for elements in an array are not as good as that for a single element HEB mixer presented in Section IV (A) in this paper. Although further work needs to be done to fully understand this discrepancy, we attribute this to several possible

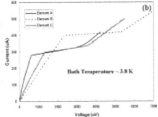
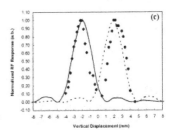

Fig. 22. Lens-antenna coupled HEB mixer imaging array characterization and results: (a) a photograph showing the components inside the close-cycled cryocooler, (b) I-V curves for HEB devices A, B and C in a one-D array, and (c) measured normalized response (dots) compared to the theoretical results by ray-tracing technique (solid- and dotted- lines).

reasons: (1) the annular slot antennas are off-axis positioned at the back side of the silicon lens and this could introduce as large as 10 dB coupling loss; (2) the array receiver system used in this research is not optimized. A four-way power combiner is utilized introducing $\sim$ 6-dB loss, which further increases the receiver noise level; (3) the array assembly strategy is designed for prototype demonstration. Wire-bonding technique is used for electrical connection, which unavoidably deteriorates the IF performance of the mixer array circuits.

4. Conclusion

In this chapter, the basic theory and mechanism for lens-antenna coupled HEB THz mixers have been introduced followed by our work on 585 GHz HEB mixers and mixer arrays based on lens-coupled ASAs. In conclusion, we have designed and fabricated hot-electron bolometer mixers, for THz heterodyne detection and imaging. The HEBs are integrated into lens-coupled annular-slot antennas that incorporate low-pass filters. DC and RF characterizations have been performed and a mixer conversion gain of -11.9 dB and a DSB receiver noise temperature of $\sim$650K have been achieved. On the basis of the single element mixer design, prototype mixer focal-plane arrays comprised of ASAs integrated with hot-electron bolometers have been developed for imaging applications at 585 GHz. Both EMF analysis and ADS momentum simulations have been performed to study the self- and mutual- impedances of the annular slot array with various element spacings. The element antenna off-axis radiation patterns have been calculated using the ray-tracing technique and the imaging angular resolution has been predicted. Initial imaging experiment results are presented and show excellent agreement with theory and simulation data, demonstrating that a diffraction-limited imaging with a resolution of $\sim$2.75 mm has been achieved at 585 GHz. Y-factor measurements show that DSB mixer noise temperatures of 1675 K and 3517 K, with mixer conversion gain of -14.73 dB and -17.74 dB, respectively, have been obtained for two adjacent elements in a one-D focal-plane mixer array. The lens-antenna coupled HEB mixer focal plane array described in this chapter provides an promising architecture for developing focal-plane arrays that capable of diffraction-limited imaging in the THz region.

5. References

[1] D. W. Floet, "Hotspot Mixing in THz Nibium Superconducting Hot Electron Bolometer Mixers," *Ph.D. Dissertation, Delft University of Technology*, September 2001.
[2] R. B. Bass, "Hot-electron bolometers on ultra-thin silicon chips with beam leads for a 585 GHz receiver," *Proceedings of the IEEE*, vol. 80, no. 11, pp. 1662-1678, November 1992.
[3] C. E. Groppi, and J. H. Kawamura, "Coherent detector arrays for terahertz astrophysics applications," *IEEE Trans. Terahertz Sci. Tech.*, vol. 1, no. 1, pp. 85-96, September, 2011.
[4] C. Kulesa, "Terahertz spectroscopy for astronomy: from comets to cosmology," *IEEE Trans. Terahertz Sci. Tech.*, vol. 1, no. 1, pp. 232-240, September, 2011.
[5] J. W. Waters, "Submillimeter-wavelength heterodyne spectroscopy and remote sensing of the upper atmosphere," *Proceedings of the IEEE*, vol. 80, no. 11, pp. 1679-1701, 1992.
[6] M. Chipperfield, "Satellite maps ozone destroyer," *Nature*, vol. 362, no. 15, pp. 592, December 1988.
[7] N. C. Luhmann, Jr., "Instrumentation and techniques for plasma diagnostics: An overview," *Infrared and Millimeter Waves*, vol. 2, , pp. 1-65, K. J. Button, Ed. New York: Academic, 1979.

5

Noise Limitations of Miniature Thermistors and Bolometers

Béla Szentpáli

Hungarian Academy of Sciences, Research Institute for Technical Physics and Materials Science, Hungary

1. Introduction

The miniature thermal resistors are comprehensively applied as thermistors or bolometers. Due to the dynamic development of the micromachining technologies became possible their mass-production with the precision and reproducibility of the microelectronics. These techniques result in typical lateral dimensions in the 1....100 μm range and thicknesses about μm, or less. The miniature devices fulfil the present-day requirements of the measuring and regulating systems demanding a large number of high-speed sensors for following quick changes, or for ensuring the quick read-out in integrated systems comprising many devices. However the higher working speed demands electronic processing circuits with broader bandwidth and consequently higher noise. Therefore it seems worth to reconsider the electronic noises of the miniature devices.

The phenomenological thermodynamic parameters depend on the average of the chaotic motion of the atoms and/or molecules. Therefore some fluctuations of their values are expected especially in very small volumes. This fact sets a physical limitation to the miniaturization; the size of the device should be large enough for representing the thermal parameters with the desired accuracy. According to the statistical physics (see eg. Kingston 1978) the mean square fluctuations of the energy is

$$\overline{\delta E^2} = kCT^2 ,$$ (1)

where k, C and T are the Boltzmann constant, the heat capacitance of the volume under discussion and the absolute temperature respectively. Because $E = CT$, (1) can be rearranged:

$$\frac{\overline{\delta T^2}}{T^2} = \frac{k}{C} \rightarrow \frac{\delta T}{T} = \sqrt{\frac{k}{C}}$$ (2)

This criterion stands a lower limit to the dimensions of the miniature thermal sensors, but it is in the nanometre region. For example in 1 μm³ platinum δT/T = 1,6*10⁻⁶ and similarly in 1 μm³ Si δT/T = 2*10⁻⁶. It depends on the application whether it is a practical limitation or not.

The temperature changes of the thermistor lead to changes of the resistance, as

$$r(T) = r_m(1 + \alpha(T - T_m)) = r_m(1 + \alpha\Delta T) ,$$ (3)

where r and r_m are the electric resistance of the thermistor at temperatures T and T_m respectively. If the thermistor is driven by a constant current (i), then the temperature can be deduced from the voltage measured on it:

$$\Delta T = \frac{1}{\alpha} \cdot \frac{\Delta r}{r_m} = \frac{1}{\alpha} \cdot \frac{\Delta U}{U_0},$$

(4)

where $U_0 = r_m \cdot i$ is the voltage on the thermistor, when its temperature is T_m.

In this chapter the limitations of the achievable accuracy and resolution by the electronic noises are treated. These noises stand an ultimate limit of the performance besides the physical limitations. The three components of the resistor noises are considered: the thermal noise, the 1/f, or flicker noise and the generation-recombination noise occurring in semiconductors. The noise induced uncertainty monotonically increases with the bandwidth; therefore it seems worthwhile to reconsider this issue for the case of fast, miniature thermal resistors. In this chapter the electronic noise is considered in a bandwidth equal to the reciprocal value of the time constant of the heat relaxation:

$$\Delta f = \frac{1}{\tau},$$

(5)

where Δf is the electronic bandwidth and τ is the characteristic relaxation time for the temperature changes. It was pointed out (Szentpáli, 2007) that the electronic processing having this bandwidth is a reasonable trade-off; such an electronics follow the $exp(-t/\tau)$ time relaxation with an accuracy of 5..15%. The accuracy improves approximately proportional with increasing bandwidth, but it grows worse quickly in narrower bandwidths.

2. Thermal model

Regarding the principal application the thermistors and the bolometers should be distinguished. The thermal equivalent circuits of them are shown in Fig.1.

In the thermistor case the thermal resistor is connected to a thermal reservoir, i.e. to a medium having much larger heat capacity than the probe itself. There is a thermal resistance between the reservoir and the thermal resistor. The thermal equivalent circuit of this situation and also the used notations is depicted in Fig. 1.a. The temperature of the reservoir holds the information on the interested physical quantity; therefore the aim is the precise measurement of T. In some applications the thermal resistance R is connected to the physical quantity, e.g. miniature Pirani type vacuum sensors (Berlicki, 2001), or similarly the thermal transfer between a heater and a temperature probe measures the flow rate of a gas, or liquid (Fürjes et al., 2004).

In equilibrium the temperature of the thermal probe, T_m is:

$$T_m = T - R\frac{T - T_A}{R + R_p} + R\frac{R_p}{R_p + R}P_i,$$

(6)

or

$$T - T_m = \frac{R}{R + R_p}(T - T_A - R_p P_i) \tag{7}$$

When T has a step–like change with $\Delta T \ll T$, then the new equilibrium $T_m + \Delta T_m$ will set in exponentially:

$$T_m + \Delta T_m = T_m + (\Delta T - R\frac{\Delta T}{R + R_p}) \cdot (1 - e^{\frac{-t}{\tau_1}}), \tag{8}$$

where

$$\tau_1 = \frac{RR_p(C + C_p)}{R + R_p} = \frac{RR_p C^*}{R + R_p} \tag{9}$$

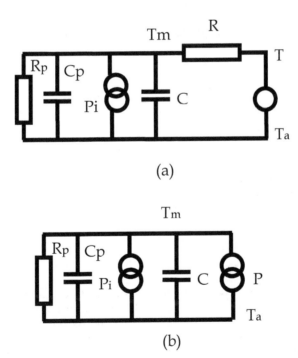

(a)

(b)

Fig. 1. The thermal equivalent circuits. (a) the thermistor configuration: the probe is connected to a thermal reservoir at temperature T; (b) the bolometer configuration: the probe absorbs the power P. Ta, Tm, C, Pi, Rp and Cp are the temperature of the ambient, the temperature of the probe, the thermal capacity of the probe, the power due to the read-out current, the thermal resistance and capacitance of the leads respectively.

For an ideal thermal probe R=0, and therefore $T_m=T$, $\tau_1=0$. Further in this ideal case the Joule heating do not changes the T_m, so the read-out current on the resistor is not limited. Of course this configuration is not accomplishable. However, realistic measurements can be made only if $R<<R_p$. In this case:

$$\tau_1 \approx RC^* \tag{10}$$

It is obvious from (7) and (8) that the effect of the wires R_p, is reduced also if T_A is close to T. This can be achieved by thermalizing the leads and supports to a temperature close to T (Berlicki, 2001). The instant temperature can be observed only if τ_1 is small compared to the speed of the temperature changes.

The other configuration is the bolometer. In this case the measured quantity is power, which absorbs in the body of the thermal resistor. The equivalent circuit model is shown in Fig.1.b. Examples are: catalytic gas detectors (Bársony et al., 2004; Barocini et. al., 2004), radiation detectors (Almarsi et al. 2006; Graf et al. 2007). In equilibrium:

$$T_m = T_a + R_p(P + P_i) \tag{11}$$

If P has a step-like change with ΔP, then the new equilibrium $T_m + \Delta T_m$ will develop exponentially:

$$\Delta T_m(t) = R_p \Delta P(1 - e^{-\frac{t}{\tau_2}}), \tag{12}$$

where

$$\tau_2 = R_p(C + C_p) = R_p C^* \tag{13}$$

Other possible question is the response to a short power pulse which duration is small compared to τ_2. In this case:

$$\Delta T_m = \frac{\Delta E}{C^*} e^{\frac{-t}{\tau_2}}, \tag{14}$$

where ΔE is the total energy of the pulse.

The heat impedance of the losses trough the leads and supports are parasitic in the thermistor arrangement; however in the case of the bolometer configuration the role of R_p and C^* are essential. A larger value of R_p increases the sensitivity; at the same time C^* should be kept small for preserving the speed and the sensitivity for short pulses. Thin wires perform these two requirements simultaneously, however the electric resistance of them should be taken into account, or more wires (4 or 3) should be applied. Also the Joule heat in the thin current wires should be managed.

Albeit the equivalent circuit in Fig. 1. is the simplified description of the real structure, the R and C values basically can be calculated numerically from the parameters of the constructing materials and the geometry. In the case of extended dimensions (leads, supports) the calculations can be made on the basis of distributed network. Beyond the "a priori" method these parameters can be deduced from measurements too. E.g. R could be determined as a limiting value of measurement series performed with successively increasing heat isolations on the probe. In general both τ values can be determined from the relaxation after a step like increment of P_i which occur remarkable heating up (Imran & Bhattacharyya, 2005).

In many cases the expressions in the frequency domain are more practical, instead of the time domain. The power spectral density of the exponential relaxation is the Lorentzian (see. e.g. Fodor, 1965):

$$e^{\frac{-t}{\tau}} \rightarrow \frac{\tau}{1+(\omega\tau)^2} \tag{15}$$

For example the power spectral density of the temperature fluctuations described by (2) is:

$$\delta\left(\Delta T^2\right)_f = \frac{4kT^2R}{1+\left(2\pi fCR\right)^2} , \tag{16}$$

where R is the heat resistance which connect the observed volume to the ambient. C is the heat capacitance of the volume under discussion, see (1), and (2). The relaxation time is τ=RC. Sometimes this noise is called "phonon noise", or fluctuation of the heat conductance.

If the heat capacitance and resistance cannot be considered as discreet elements, but distributed parameters along one dimension, then (Socher at al., 1998):

$$\delta(\Delta T^2)_f = \frac{16kT^2R'}{\pi L(R'G' + \frac{\pi^2}{4L^2})} \frac{1}{1+(\frac{2\pi fR'C'}{R'G' + \frac{\pi^2}{4L^2}})^2} , \tag{17}$$

where R'and C' are the heat resistance and the heat capacitance per unit length; G' is the heat conductance to the ambient per unit length and L is the length of the structure, considered as being one dimensional.

3. The thermal noise

The Nyquist, or Johnson noise is generated by the thermal motion of the electrons. This is a "white" noise, i.e. the noise power in unity bandwidth is constant, independent on the frequency. The cut off frequency at room temperature is in the terahertz region (Hooge et al., 1981). It occur voltage, or current fluctuations depending on the embedding circuit (van der Ziel, 1986). The variance of the voltage across the resistor terminals is:

$$\overline{\left(U-\overline{U}\right)^2} = \delta U^2 = 4kTr\Delta f , \tag{18}$$

where Δf is the bandwidth in which the noise is measured. The square root of the variance is the standard deviation, which is the so called "thermal voltage":

$$\delta U = \sqrt{4kTr\Delta f} \tag{19}$$

This voltage fluctuation is present on the resistor even without bias. Therefore the relative importance of the thermal noise is smaller when the useful signal increases, i.e. when the bias on the resistor is large. However, the bias is limited by the self-heating of the thermal resistor.

In the "thermistor" case an upper limit of the bias can be derived from the accuracy. The Joule heating results in temperature drop on R. The maximum of this drop can be prescribed as:

$$i^2 rR \leq pTm \, , \tag{20}$$

where i is the biasing current and p is constant, which express the required accuracy. From (6) and (20) follows the maximum of the voltage drop on the resistor r:

$$U_0 = \sqrt{\frac{pr(R + R_p)T_m}{RR_p}} \tag{21}$$

Substituting (19) and (21) into (4) and taking into account (10) it gives:

$$\delta T_{th} = \frac{1}{\alpha} \frac{\delta U_{th}}{U_0} = \frac{1}{\alpha} \frac{\sqrt{4kTr / \tau_1}}{U_0} = \frac{2}{\alpha} \sqrt{\frac{kT}{pC^* T}} \tag{22}$$

This is the uncertainty of the temperature read-out due to the thermal noise if the bandwidth is $1/\tau_1$. It is worthy to note that this uncertainty does not depend on the actual value of r and R. Except the constants it depends only on the ratio of the thermal energy and the heat accumulated in the heat capacitance, C^*.

For bolometer case the rearrangement (11) and (4) gives:

$$\delta P_{th} = \frac{\delta T_{th}}{R_p} = \frac{1}{\alpha R_p} \frac{\delta U}{U} = \frac{1}{\alpha R_p} \frac{\sqrt{4kT_m r / \tau_2}}{ir} = \frac{2\sqrt{kT_m}}{i\alpha R_p^{3/2} \sqrt{rC^*}} \tag{23}$$

In these measurements there is no need for accurate determination of temperature, therefore the choice of the bias current and the resistor temperature T_m depends on different considerations. E.g. the catalytic gas detectors work at temperatures about 700 °C, which is ensured by the bias current (Fürjes et al., 2002).

In the similar way from (14) it gives:

$$\delta E_{th} = \delta T_{th} C^* = \frac{C^*}{\alpha} \frac{\delta U}{U} = \frac{C^*}{\alpha} \frac{\sqrt{4kT_m r}}{ir\sqrt{R_p C^*}} = \frac{2\sqrt{kT_m C^*}}{i\alpha \sqrt{rR_p}} \tag{24}$$

For small incident powers, or energies T_m in (23) and (24) depends on the read out current. $T_m = T_0 + R_p i^2 r$. At elevated temperatures, when $T_m \gg T_0$, the δP_{th} and δE_{th} become independent on the current according to (23) and (24). However, it should be noted that at elevated temperatures even some deterioration of accuracy can occur, because the value of α decreases at high temperatures for the greatest number of materials. The increase of the current in the region, where it gives rise to only moderate temperature increment improves the accuracy.

4. The flicker (1/f) noise

The flicker or 1/f noise is also a general component in all resistors. It originates from the fluctuation of the resistance, concretely from the fluctuations of the mobility, or in other words from the fluctuations of the thermal scattering of mobile charge carriers (Kogan, 1996). Therefore this effect stands an absolute limit to the accuracy of the measurement; it can not overcome by increasing the bias. The spectrum of the 1/f noise is:

$$\frac{\delta r_f^2}{r^2} = \frac{C_{1/f}}{f} \tag{25}$$

The dimensionless number $C_{1/f}$ is the measure of the magnitude of the noise, δr_f^2 is the spectral density of the variance of the resistance, i.e. the variance measured in unity bandwidth. $C_{1/f}$ makes possible to compare the noise levels observed under different conditions as frequency ranges, current or voltage. In metallic and semiconductor resistors the Hooge-relation (Hooge, 1969) is valid:

$$C_{1/f} = \frac{\alpha_H}{N}, \tag{26}$$

where α_H is the so called Hooge-constant and N is the total number of mobile charge carriers in the resistor. The value of α_H is not universal constant, as it was supposed earlier, when Hooge discovered the above relationship. It varies from about 0.1 to 10^{-8} for different materials and structures (Kogan, 1996). Largest values of α_H were obtained in strongly disordered and inhomogeneous conductors, e.g in high-Tc superconductors, as it will be presented in section 9. As $C_{1/f}$ is able to compare different measurements, the Eq. (26) offers more general facility of comparison, because α_H is a specific material parameter, independent of the number of charge carriers, i.e. the volume of the resistor. Therefore not only the noises obtained under different measuring circumstances on equivalent resistors can be compared, but also noises of samples having different sizes. On other hand Eq. (26) shows that the noise is larger in smaller resistors made from the same material. It should be noted here that there are other 1/f noises too; e.g. in the channel of MOS transistors the 1/f noise is produced by fluctuation in number of electrons trapped in the oxide (Kingston (Ed.), 1957).

It should be realized that the spectrum cannot be exactly 1/f in the whole frequency range from zero to infinity. At first the spectrum is undefined at f=0, because the zero divider in (25). Further the total noise power, which is the integral of the spectrum is the logarithm function, which has infinite values when f→0 and when f→∞. Therefore it is generally assumed, that the spectrum flattens below a certain low frequency and it should be stepper over a certain high frequency. However, neither of this frequency limits have been observed yet. In spite of this fact the integration of the spectrum can be performed in a finite frequency region, which does not involve the 0 Hz (Szentpáli, 2007). The variance of the resistance is taken as:

$$\delta r^2 = r^2 C_{1/f} \int_{f_1}^{f_2} \frac{df}{f} = r^2 C_{1/f} \log_e\left(\frac{f_2}{f_1}\right),$$ (27)

where $f_2 = 1/\tau$ and $f_1 = f_2/a$, and expediently $a \gg 1$. The relative fluctuation of the resistance is:

$$\frac{\delta r}{r} = \sqrt{C_{1/f} \log_e a}$$ (28)

It should be noted here that in this treatment of the $1/f$ spectrum that the noise in the bandwidth is independent on the frequency, it depends only on the ratio a. However due to the square root and log function, this dependence is rather weak. For sake of simplicity in this paper $a = 10^6$ will be used. It means that

$$\frac{\delta r}{r} = 3.7\sqrt{C_{1/f}}$$ (29)

As it was mentioned the choice of the "a" value is not decisive, e.g. taking only 3 decade bandwidth the multiplayer in (29) would be 2.6. Similarly 9 decade bandwidth results in a numeric factor of 4.5. Even the value of $C_{1/f}$ is not known with better accuracy. Suggestively the value of f_1 can be interpreted as the reciprocal of the time of observation. A numeric example: $a = 10^6$ can mean that $f_2 = 10$ kHz is the upper cut-off frequency of the electronics belonging to 0.1 ms relaxation and $1/f_1 = 100$s is the time of the measuring.

The fact that the fluctuation due to the flicker noise is hardly sensitive to the bandwidth and unrelated to the absolute value of the frequency means that it is not worth to limit the speed of the electric circuit.

Substituting (29) into (4) we obtain the $1/f$ noise equivalent uncertainty of the temperature of the thermistor:

$$\delta T_{1/f} = \frac{3.7\sqrt{C_{1/f}}}{\alpha} = \frac{3.7\sqrt{\alpha_H}}{\alpha\sqrt{N}} = \frac{3.7\sqrt{\alpha_H}}{\alpha\sqrt{nV}},$$ (30)

where n is the volume density of free charge carriers and V is the volume.

For the bolometer configuration it gives

$$\delta P_{1/f} = \frac{\delta T}{R_p} = \frac{1}{R_p}\frac{3.7\sqrt{\alpha_H}}{\alpha\sqrt{nV}}$$ (31)

The noise equivalent pulse uncertainty is:

$$\delta E_{1/f} = \delta T C^* = C^*\frac{3.7\sqrt{\alpha_H}}{\alpha\sqrt{nV}}$$ (32)

Here it is worthy of note that the introduction of the integral (27) makes possible to express the figures of merit as δT, δP and δE on the basis of (26), which is a principal physical relation. The general conclusion on the $1/f$ noise is that the measuring circumstances practically cannot influence it. It can be influenced only by the construction, by the α_H value of the applied material and the volume. The increase of the volume decreases $\delta T_{1/f}$ and $\delta P_{1/f}$ at the sacrifice of the speed. $\delta E_{1/f}$ is proportional to $C^* \sim V$, the $1/f$ uncertainty changes as $V^{-1/2}$, therefore finally $\delta E_{1/f}$ is proportional to $V^{1/2}$. In this case the smaller volume is clearly advantageous.

5. The generation-recombination noise

In semiconductor resistances the generation-recombination of mobile charge carriers from trap states give rise also to the fluctuations of the resistance. This type of noise is absent in metals, moreover plays no role in the present-day silicon material, which is practically free from deep centres. The probability of the emission from the trap state is exponential, while the recombination is obeyed by the mass-action law. The simplest case of the generation-recombination process is when there is only one trap centre, and then the related noise spectrum is the transformed exponential function: the Lorentzian, (Jones, 1994). This is shown in Figure 2 together with spectra of the other noises, mentioned earlier.

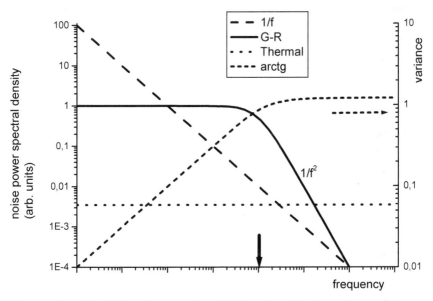

Fig. 2. The spectra of the different noises. The right hand scale is the variance of the g-r spectrum.

At low frequencies always the $1/f$ noise prevails over the other noise mechanisms, on the high frequency part of the scale the thermal noise dominates. The crossing point of the two spectra is about 1 kHz for such granular materials as the graphite and it sinks to 1 Hz region, or below for metals; for semiconductors this point is between the two values, depending on the material and the doping. The breaking point of the Lorentzian spectrum of the generation-recombination (g-r) noise is $1/\tau$; it depends exponentially on the temperature. This noise can be observed in a wide frequency range, if there is a noticeable amount of traps.

The Lorentzian spectrum of this type of resistance noise is:

$$\frac{\delta r^2}{r^2} = \frac{M\tau_{g-r}}{1+(2\pi\tau_{g-r}f)^2}, \tag{33}$$

where τ_{g-r} is the characteristic g-r relaxation time, which is the harmonic mean of the emission and capture times of the trap. M is the magnitude of the spectrum. In the case of more than one trap the situation becomes rather sophisticated, the spectrum can be the sum of the characteristic spectra, or it can be a new Lorentzian with a mixed τ_{g-r} value [Hooge , 2002, 2003). This case is out of the scope of the present discussion. The variance of r obtained in a finite bandwidth is:

$$\frac{\delta r^2}{r^2} = \int_0^f \frac{M\tau_{g-r}}{1+(2\pi\tau_{g-r}f')^2}df' = \frac{M}{2\pi} \cdot arctg(2\pi\tau_{g-r}f) \tag{34}$$

This function is also depicted in Figure 2, at frequencies $f<<1/2\pi\tau_{g-r}$ it can be approximated by linear function, while at higher frequencies it approximates the saturation value $\pi/2$. The noise equivalent uncertainties can be calculated for the two cases separately.

At low frequencies the fluctuation is:

$$\frac{\delta r}{r} = \sqrt{M\frac{\tau_{g-r}}{\tau}}, \tag{35}$$

where $1/\tau$ is the noise bandwidth, it is equal to $1/\tau_1$ or $1/\tau_2$ are for the thermistors or for the bolometer cases respectively. The corresponding uncertainties are:

in the thermistor arrangement

$$\delta T_{g-r} = \frac{1}{a}\sqrt{\frac{M\tau_{g-r}(R+R_p)}{RR_pC^*}} \cong \frac{1}{a}\sqrt{\frac{M\tau_{g-r}}{RC^*}} \tag{36}$$

in the bolometer set-up

$$\delta P_{g-r} = \frac{1}{\alpha R_p}\sqrt{M\frac{\tau_{g-r}}{\tau_b}} = \frac{1}{\alpha R_p}\sqrt{M\frac{\tau_{g-r}}{R_pC^*}} \tag{37}$$

and for the energy pulse

$$\delta E_{g-r} = \frac{1}{\alpha}\sqrt{\frac{M\tau_{g-r}C^*}{R_p}} \tag{38}$$

At frequencies $f >> 1/2\pi\tau_{g-r}$ the effect of the g-r noise saturates, because practically the whole spectrum is taken into account. In this case the uncertainties can be expressed similarly to (30), (31) and (32), only the numeric factor $3.7\sqrt{C_{1/f}}$ should be changed to $\sqrt{M}/2$.

Thus:

$$\delta T_{g-r} = \frac{\sqrt{M}}{2a} \tag{39}$$

$$\delta P_{g-r} = \frac{\sqrt{M}}{2aR_P} \tag{40}$$

$$\delta E_{g-r} = C^* \frac{\sqrt{M}}{2a} \tag{41}$$

6. The thermopile

The thermopiles are related to the bolometers; in many cases they can substitute each other. The common feature is the temperature change of the sensitive element; however the operation principle of the thermopile is based on the Seebeck effect. The sketch of a miniature thermopile fabricated by micromachining is shown in Fig. 3.

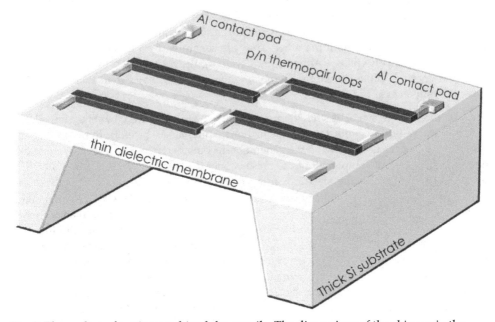

Fig. 3. The outline of a micromachined thermopile. The dimensions of the chip are in the mm range, the thickness and width of thermopair lines are typically in the sub micrometer and micrometer ranges respectively.

As it is obvious the miniature thermopile is not suitable to measure the absolute temperature; this task can be fulfilled by wire thermopairs; which is a well developed technique. The thermopiles sense the temperature difference; they can be applied for measuring the physical quantities which can be transformed to temperature difference (Graf et al., 2007). The absorbed power is such a quantity; it heats up more effectively the inner part of the thermopair loops, than the outer parts lying on the good heat conducting substrate. The strong thermal asymmetry is a key point of the operation. The central region on the membrane changes its temperature, while the outer parts lying on the good heat conducting substrate are thermalized to the ambient temperature. The serial connection of many thermopairs multiplies the output signal. Numerous different constructions are applied both as regards the materials and the geometrical construction. The micromachined thermopairs suitable can be produced by CMOS technology (Lenggenhager et al., 1992), in this case the one arm of the thermopairs is formed from p- or n- polycrystalline silicon and the other arm is aluminium in many cases. The alloys from different composition rates of Bi-Sb-Te are also popular due to their high Seebeck coefficient. Regarding the geometry there are closed membranes, as it is shown in Fig.3., beam-type membranes and a combination of the two the so called "bridge-type" membrane. These two latter constructions aim to increase the thermal resistance by reducing the contact between the membrane and the silicon rim. However the cantilever beam shaped membranes reduces the mechanical stability, and often become bent due to the different heat expansion coefficient of the multilayer structure. The bridge-type construction overcomes on this problem, the membrane is contacted on all sides to the substrate, but the contact areas are reduced. The thermopair configurations show also numerous variations. They can be arranged in parallel loops, or radial distribution. In some cases one strip lies upon the other. They are separated by a thin dielectric film; e.g. polycrystalline silicon - SiO_2 - Aluminium. Grouping the thermopairs linearly instead of loops the lines can act as miniature dipole antennas and the device detects the mm/THz waves (Szentpáli et al., 2010).

Because the output signal is the thermoelectric voltage measured by high impedance voltmeter there is no significant current in the thermopairs. Therefore the noises of the resistance – the $1/f$ fluctuation and the generation-recombination noise – play no role in this device. The important component of the noise is only the thermal noise of the resistor. The output voltage and also the electric resistance are proportional to the number of loops, however, the thermal noise voltage scales only to $\sim\sqrt{r}$, therefore the signal to noise ratio improves with increasing number of loops. If the area of the device is limited then sooner or later the number of the loops can be increased only by decreasing their widths. In this case the resistance will be proportional to the square of the number of the loops, and the signal to noise ratio will not improve, only the signal will be larger. However, it should be kept in mind that as the covering of the membrane grows the thermal resistance decreases and also the temperature difference on the thermopairs reduces in applications where the input power is fixed, e.g. radiation detectors, etc. The temperature distribution along the thermopair arms depends on the configuration and also on the excitation (Socher et al., 1998; Xu et al., 2010; Ebel et al., 1992). The common feature is the increment of the temperature difference with the length of the thermopairs. If the major heat conductance is through the thermopair strips, then this dependence is quadratic, the cooling through the atmosphere and the heat loss through radiation decreases this dependence, but it remains super-linear. Therefore the increase of the length of the thermopairs will result in the improvement of the signal to noise ratio.

While the bolometers work under isothermal conditions, the output signal is the average of the effects in the volume. Therefore the temperature fluctuation described by (2) is practically smoothed to invisible. This is not trivial for thermopairs. Here the temperature and its gradient are distributed somehow along the sensor wires and even a small section can add a determining amount to the output voltage. Therefore it is worth to investigate this effect. The spectral distribution of the thermal fluctuations has maximum when $(2\pi f CR)^2 \ll 1$. In this case (16) can be simplified:

$$\delta\left(\Delta T^2\right)_f = 4kT^2R \ [\frac{K^2}{Hz}] \tag{42}$$

The corresponding spectral distribution of the output voltage fluctuation is $(\delta U)^2 = S^2 * \delta(\Delta T^2)$, where S [V/K] is the Seebeck coefficient of the conductor. It is noted here, that the absolute Seebeck coefficient of most of the conducting materials are determined, or it can be measured by coupling them to lead, which has a low and precisely known Seebeck coefficient as a function of temperature (van Herwaarden & Sarro, 1986). The voltage fluctuation on an elementary length dl is:

$$\delta(U^2)_f = S^2 4kT^2 \frac{dl}{A\sigma_{th}}, \tag{43}$$

where A and σ_{th} are the cross section area and the thermal conductivity of the wire respectively. This fluctuation can be compared by the thermal voltage fluctuation (18) of the same elementary length:

$$\delta_{th}(U^2)_f = \frac{\delta U^2}{\Delta f} = 4kTr = 4kT\rho\frac{dl}{A}, \tag{44}$$

where ρ is the electrical specific resistance of the material. The ratio of the two noises is:

$$M = \frac{\delta(U^2)_f}{\delta_{th}(U^2)_f} = \frac{S^2T}{\sigma_{th}\rho} \tag{45}$$

The values of M are listed in Table 1. for two metals and two differently doped Si at 300 K.

	ρ [Ωm]	S [μV/K]	σ_{th} [W/mK]	M
Platinum	$1.06*10^{-7}$	4.45	71	$7.9*10^{-4}$
Aluminum	$2.65*10^{-8}$	1.6	237	$1.2*10^{-4}$
silicon p = 10^{25}m^{-3}	$1.0*10^{-4}$	644	150	$8.3*10^{-3}$
silicon n= 10^{21}m^{-3}	$4.5*10^{-2}$	1958	150	$5.7*10^{-7}$

Table 1. The ρ, S, σ and M values for platinum, aluminum and differently doped silicons.

It seems that the statistical fluctuation of the temperature has a negligible effect besides the thermal noise. The ratio M increases proportionally to the temperature, because the temperature fluctuation is proportional to T^2, while the thermal noise growth with T only. Using the data of the table 1. as a rough estimation the M=1 ratio could be achieved only at that temperatures, where the materials are already melted, or vaporized.

7. Thermal radiation

The bolometers are very much employed for sensing the thermal radiation. This topic has a large literature (see eg. Hennini & Razeghi, 2002). In this frequent case, the fluctuation in the photon flux sets a physical limit to the accuracy. The power emitted by the thermal radiation in the half space is expressed by the Stefan-Boltzmann formula, multiplied by the emissivity of the body under discussion:

$$P = A\varepsilon\sigma_{S-B}T^4,$$
(46)

where A is the emitting surface, ε is the emissivity and σ_{S-B} = 5.67*10⁻⁸ Wm⁻²K⁻⁴ is the Stefan-Boltzmann constant. The emissivity of the ideal black-body is ε =1. Realistic bodies are "grey", their emissivity is ε <1. As a thumb rule the emissivities of metals are ε~0.1...0.2 and for dielectrics ε~0.7...0.8. Still in more precise description the emissivity depends on the wavelength, $\varepsilon=\varepsilon(\lambda)$ (Kruse et al., 1962) and also on temperature. The emissivity describes also the measure of the absorption; in equilibrium every body emits and absorbs the same amount of the radiation at each frequencies. The emissivity of miniature bolometers can be improved significantly by covering with a multilayer resonant layer structure (Liddiard, 1984, 1986 and 1993). This improving is successful also for thermopairs (Roncaglia et al., 2007). For the same purpose porous gold layers, so called gold black can be applied too.

The emission of photons is a random process governed by the Einstein-Bose statistics. The spectral fluctuations of the emitted power can be derived on this base (Kruse et al., 1962) as:

$$\delta P_r = 8A\varepsilon k\sigma_{S-B}T^5$$
(47)

It is noted here that (47) describes the power spectral density in the function of the frequency of the radiation and not against the frequency of the electronics. The observed noise is the integral of (47) for the radiation transmission and detection region.

In an application there are minimum three different components: the body of which the temperature is detected, the sensor and the ambient. For this case:

$$\delta P = 8F_{(1\to2)}A_1\delta_{(S-B)}\varepsilon_1\varepsilon_2T_1^5 + 8F_{(3\to2)}A_3\varepsilon_{(S-B)}\varepsilon_3\varepsilon_2T_3^5 + 8A_2\varepsilon_{(S-B)}\varepsilon_2T_2^5,$$
(48)

where $F_{i\to j}$ are the view factors and the indexes 1, 2, 3 denotes the observed body, the sensor and the ambient respectively. In radiative heat transfer, a view factor $F_{i\to j}$, is the proportion of all that radiation which leaves surface i and strikes surface j (Lienhard IV & Lienhard V, 2003). Eq. (48) expresses the ultimate limit of all thermal radiation detectors. In the practical arrangement the sensor "views" mainly the observed object, therefore the first term in the right side of (48) is the greatest. This fact limits the possibility of decreasing the noise by cooling the sensor and/or the ambient.

8. Figures of merit

The δT_{th}, δP_{th} and δE_{th} means those values which are required to produce voltage on the thermistor equivalent to the electronic noise. If all the three mentioned noise mechanisms are significant as it is shown in Fig. 2., then the resulting uncertainty contribution should be calculated as the square root of the sum of variances of the independent fluctuations, e.g:

$$\delta P = \sqrt{\delta P_{th}^2 + \delta P_{1/f}^2 + \delta P_{g-r}^2} \qquad (49)$$

and similar relation for δT and δE. Sometimes these values are called "nominal detectable signal", or noise-equivalent signal. The minimum detectable quantities are larger for two reasons: the significantly detectable amount should be over the noise floor at least by a factor of 2, or rather 2.5 if the definition of the tangential sensitivity (H.A. Watson, 1969) is applied; further due to the finite bandwidth the detected signal is not the instant peak value, it is attenuated to a certain degree.

The expressions are not completely unambiguous in the literature. Usually the term "sensitivity" is used for the temperature measurement and the phrase "responsivity" for the power sensing. This later is the input–output gain of a detector system, its dimension is V/W, or A/W depending on the output signal.

A further parameter is the bandwidth of the system. In this chapter the calculations were performed for the greatest rational bandwidth, specified by the temperature settling time. In the ordinary way the noise equivalent power/ temperature/energy is expressed for a 1 Hz bandwidth and the speed of the sensor is given independently.

For radiation detectors the "specific detectivity" is a widespread figure of merit. It is equal to the reciprocal of noise equivalent power (*NEP*), normalized to unit area and unit bandwidth:

$$D^* = \frac{\sqrt{A * \Delta f}}{NEP_{(1Hz)}}, \qquad (50)$$

where A is the area of the photosensitive region of the detector, Δf is the effective noise bandwidth, and the NEP is the noise equivalent power in unit bandwidth. Of course this definition is unambiguous only in the case of "white noises", as the thermal noise; for the „colour noises" -the $1/f$ and the g-r noise - also the absolute value applied frequency should be defined. The common units of D* are $cmHz^{1/2}/W$. D* also called the Jones in honour to R. Clark Jones who defined this magnitude (Jones, 1949).

9. Superconducting bolometers

The resistance of the superconductors performs a very sharp change in the transition region. The transition temperature (Tc) of the superconducting metals fall in the cryogenic temperature region mostly at around, or above 1 K, for some special alloy it can be around 10 K. E.g. for $Mo_{0.6}Re_{0.4}$ the transition temperature is 12.6 K (McMillan, 1968). There is an other group of materials, the so-called high-Tc superconductors, where Tc is in the 80...130

K range; above the boiling point of liquid nitrogen (77 K). These compound materials consist of 4-6 different components, their crystal structure is typically tetragonal. The transition temperature depend also on the magnetic field and through that on the current flowing in the bolometer. Fig. 4. shows the resistance change in the transition region.

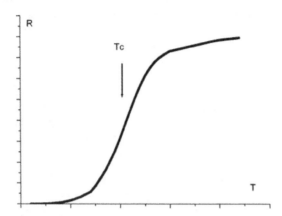

Fig. 4. The resistance in the transition region of superconductors. The axis are in arbitrary units.

The transition region is a few K for the high-Tc superconductors and only a few hundredths K for superconductors having transition temperature in the 1 K region. In the transition region the α is very high, it can overcome even the value of 5 (de Nivelle et al., 1997). These cooled bolometers are applied for sensing radiations; other applications fit hardly to the cryogenic surrounding. The setting into operation of the high-Tc bolometers is relative easy, because they can be cooled with liquid nitrogen or small cryocoolers. The cooling to around 1K , or even deeper needs big, sophisticated and expensive techniques. For both group of superconductors the effect of electrothermal feedback should be taken into account. The origin of this effect is the high value of α in the transition region. The functioning of the electrothermal feedback is shown in Fig. 5.

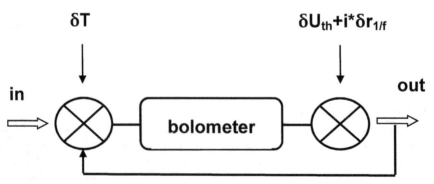

Fig. 5. The sketch of the electrothermal feedback, after de Nivelle et al., 1997.

The input radiation heats up the resistor therefore its resistance and voltage on it will increase if it is supplied by constant current. The higher voltage causes further dissipation and consequently heating. The feedback loop in the Fig. 5. takes account of this effect; the increment of the electric power dissipation adds up to the input signal. The gain of this loop is:

$$L = (P\alpha R_p) / \sqrt{1 + (2\pi\tau_e)^2} = L_0 / \sqrt{1 + (2\pi f\tau_e)^2} , \qquad (51)$$

where P is the absorbed radiating power and $\tau_e = \tau_2/(1 - L_0)$. The responsivity of the device is:

$$RES = \frac{1}{i} \frac{L_0}{1 - L_0} \frac{1}{\sqrt{1 + (2\pi f\tau_e)^2}} \quad [\frac{V}{W}] \qquad (52)$$

For stable operation L_0 should be smaller than 1. In the literature $L_0 = 0.3$ is selected as an optimal value. If the bolometer is biased by constant voltage then L and RES become negative. In this case L_0 should be larger than -1 (de Nivelle et al., 1997).

The statistical fluctuation of the temperature (16) adds to the input, the equivalent power is:

$$P_{\delta T} = \frac{1}{R} \sqrt{\frac{4kT^2 R}{1 + (2\pi fCR)^2}} = \sqrt{\frac{4kT^2}{R(1 + (2\pi fCR)^2)}} \qquad (53)$$

The corresponding spectral density of the voltage noise power:

$$\delta U_{\delta T}^2 = RES^2 \frac{4kT^2}{R(1 + (2\pi fCR)^2)} \quad [\frac{V^2}{Hz}] \qquad (54)$$

The thermal noise voltage (19) is added expediently to the voltage on the resistor. This increases the excess power dissipation by δV_{th} is: $\delta V_{th}i$. Finally, the thermal power spectrum of the thermal noise:

$$\delta U_{th}^2 = 4kTr(1 + iRES) \qquad (55)$$

The 1/f noise is the fluctuation of the resistance, see (25). It adds to the voltage $\delta r*I$. The power spectrum can be derived similar as above:

$$\delta U_{(1/f)}^2 = (\frac{riC_{(1/f)}}{f})^2(1 + iRES) \qquad (56)$$

The G-R noise is also the fluctuation of the resistance, it could be treated similarly, but it plays no role in the superconductors, at least there is no mention on it in the literature.

The $C_{1/f}$ value of metals and simple alloys is very small due to the enormously large concentration of the mobile electrons. Therefore the 1/f noise is practically absent in bolometers having transition temperature in the 1...10 K range. However, the statistical fluctuation of the temperature can be observed (Maul et al., 1969).

The 1/f noise is significant in the high-Tc superconductors. The α_H values range from $5*10^{-4}$ to $1.4*10^3$ for different material compositions (Khrebtov, 2002). The epitaxial growth of the sensitive layer decrease the 1/f noise (de Nivelle et al. 1997), however, the α_H values show rather large scatter between the samples prepared in the same run, further α_H depends also on the α value in the same sample, i.e. it depends on the workpoint within the transition region. The noise spectrum fits to the $\sim 1/f^\beta$ characteristics with $\beta \sim 1$ from room temperature to the transition region; in the transition region the value of β changes between 0.8 and 2 (Khrebtov, 2002). In spite of these uncertainties the observation of the statistical fluctuations of the temperature was successful (de Nivelle et al., 1997). In general the high-Tc superconducting bolometers perform a remarkable progress as the improved technologies decrease the excess 1/f noise. The first published structure had $\alpha_H \sim 10^5$ and $D^* \sim 10^7$ cmHz$^{1/2}$/W, while in 2002 the best published data was $D^* = 1.8*10^{10}$ cmHz$^{1/2}$/W (Khrebtov, 2002).

10. Numeric examples

10.1 The pellistor

The first example is the "unsupported" pellistor, described in (Fürjes et al., 2002). It is applied for detection of combusting gases. This is a meander shaped resistor prepared from sputtered Pt film on the top of a SiO_2-SiN_x double layer from beneath the silicon substrate was removed (Dücső et al., 1997); see the photo in Fig. 6.a. It is used in bolometer regime, 18 mW electric input power heats it up to 780K. The heat capacitance is 41.57 nJ/K, the heat resistance to the surroundings is $R_p = 26.9$ K/mW, $\tau_2 = 1.15$ ms. The electric resistance at 780 K is about 411 Ω and the measured thermal coefficient of the resistance $\alpha = 6.63*10^{-4}$ K^{-1}. The substitution of these data into (23) and (24) gives $\delta P_{th} = 2.6$ nW and $\delta E_{th} = 2.9$ pJ.

The effect of the 1/f noise can be taken into account by means of the $C_{1/f} = \alpha_H/N$ parameter. The measured α_H parameters of sputtered Pt films range from 10^{-4} to $2*10^{-3}$ (Fleetwood & Giordano, 1983). The published α_H values are related to the number of atoms instead of the number of conducting electrons. In this way the problems of the rather complex Fermi surfaces are skipped. The mass of this resistor is $2.33*10^{-8}$g. These data with the largest published α_H give $C_{1/f} = 2.8*10^{-17}$. The upper frequency limit of the bandwidth in the present case is in the kHz range; the integration in a bandwidth of six decades means that the lower limit falls in the mHz range. After substitution (31) and (32) give $\delta_{1/f}P = 1$ nW and $\delta_{1/f}E = 1.2$ pJ respectively.

The resulting noise: $\delta P = \sqrt{\delta P_{th}^2 + \delta P_{1/f}^2} = 2.8 nW$, and $\delta E = \sqrt{\delta E_{th}^2 + \delta E_{1/f}^2} = 3.1 pJ$, which are scarcely greater than the thermal noise, consequently even the pessimistically estimated 1/f noise plays no significant role under these circumstances. When the lateral dimensions of the device decrease by a factor of ten and the thickness of the platinum remains unchanged the electric resistance does not change and the volume of the resistor decreases by a factor of 100. In this case both δP_{th} and $\delta P_{1/f}$ increases ten times due to the decrease of C^* and increase of $C_{1/f}$.

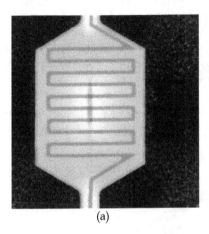
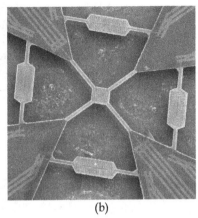

(a)   (b)

Fig. 6. (a) The photo of the gas sensor pellistor, the size is 100x100 μm^2; (b) the photo of the flow-meter chip.

10.2 The flow sensor

The photo of the device is shown in Fig.6.b. It consists of 4 bolometers described above and a central heater fabricated with the same technique. The gas (air) delivers the heat from the heater to the sensors by conductive and convective mechanisms. The heat transfer is equal to a heat resistance R~2 K/mW, at the flow velocity of 1 m/s. It is more than ten times less than R_p, ; the τ_1 ~ 85 μs. The medium working temperature is 310 K, the electrical resistance 289 Ω, the thermal coefficient of the electrical resistance at this temperature is α=2.3*10^{-3} and the estimated value of p=10^{-2}. It is noted here, that the α value for sputtered metal layers are lower than that for well heat treated wires. From these data $\delta T_{th} \cong 0.88\ mK$. Substituting the above data in (30) the value of $\delta T_{1/f}$=0.029 mK is obtained. Also in this case The decrements of the lateral dimensions result in the simultaneous increase of both noises.

10.3 The implanted silicon resistor

The third example is an ion implanted Si resistor, measuring the temperature of the chip containing piezoresistive pressure sensors (Szentpáli et al., 2005). The parameters of the thermal resistor are not optimized for thermistor function; it was fabricated simultaneously with the piezoresistors, with the same B implantation step. The layout of the resistor is U shaped: two 150 μm long arms connected with a 40 μm bottom part. The width is 20 μm. The resulting doping profile is Gaussian, with a surface concentration of 6*10^{18} cm^{-3} and 2.3 μm depths. The resistors have a room temperature resistance of 2.3 kΩ, the temperature coefficient α=1.6*10^{-3} K^{-1}. The heat capacitance can be calculated from the geometrical data, it is C=25.5 nJ/K. The heat resistance to the chip can be estimated from the electric spreading resistance. Between the p-type resistor and the n-type substrate there is a p-n junction. At large forward biases the I-V characteristics of the junction declines from the exponential due to the serial resistance. In the present case this resistance is 21Ω, and the analogous heat resistance is about 5 K/W. The very low value of the heat resistance allow very quick work, τ_1=0.12 µs. (This is the reason of this indirect estimation of R, because a measurable heat relaxation would need too high heat pulse and precise resistance measurements at about 10 MHz bandwidth.) The connections between the resistor and the bonding pads are evaporated Al strips. Their temperature is equal to temperature of the chip, so the situation is similar to the previous one; the heat transfer from the thermistor to the leads is negligible. Supposing a 1 mA bias and the parameters used above (22) gives δT_{th}= 5 mK. (23) and (24) results in δP_{th}=1 mW and δE=127.5 pJ respectively. This device is a precise and very fast thermistor; however, the bolometer performance is limited by the low heat resistance R.

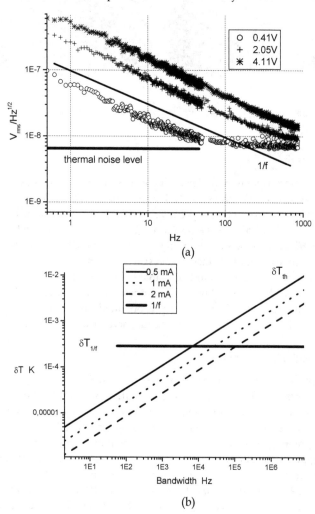

Fig. 7. (a) The measured noise spectra of the ion-implanted resistor. The continuous line stands for the 1/f characteristics; (b) Calculated δT_{th} and $\delta T_{1/f}$ at different biases and bandwidths.

The 1/f noise spectra of the resistor are shown in Fig.7.a. The spectra shift proportional to the bias, proving that the noise is caused by resistance fluctuations. From the spectra $C_{1/f}$ = 1.6*10^{-14} is obtained. Substituting this value in (30) $\delta T_{1/f}$ = 0.29mK, which is negligible comparing to the thermal noise. However most of the applications do not need so broad bandwidth, which is enabled by the thermal relaxations. If the electronic bandwidth is tighter, then δT_{th} will be smaller too and becomes comparable to the $\delta T_{1/f}$ or even sinks below it. In other words: the 1/f noise is the minimum of the total noise, which can be reached only in tight bandwidths. This is shown in Fig. 7.b.

11. Conclusions

The noise limited sensitivities of miniature thermal resistors were calculated. The statistical fluctuation of the temperature determines the lower limit of their size. The thermometer-type and the bolometer-type thermistor configurations were considered each with three different physical noise mechanisms: the thermal, 1/f and generation-recombination noises respectively. Special attention was put to the speeds of the measurements; which were limited by the thermal relaxations. The noise equivalent signals as δT in the thermometer configuration, δP and δE in the bolometer arrangement were calculated in the bandwidths limited by the thermal relaxations. It was shown, that under this circumstances $\delta T \sim \sqrt{k/C^*}$ and independent on the values of electric resistance and heat conductance. A method was proposed for the calculation of the noise equivalent signals from the spectra of the 1/f noise. The fluctuation in the thermal radiation set a physical limitation to the attainable accuracy, this effect was also treated. The thermopiles are closely related to the bolometers, it was shown that the main noise component in them is the thermal noise; they are free from the resistance fluctuations and in the practical cases the thermal noise exceeds the noise from the statistical fluctuation of the temperature. In the transition region of the superconducting bolometers the electrothermal feedback occurs due to the great thermal coefficient. The statistical fluctuation of the temperature is the main limitation of the performance of devices having Tc at deep temperature. The high-Tc superconductors show large 1/f noise, however they improve significantly with the development of the fabrication techniques. The statistical fluctuation of the temperature was observed even with such a bolometer. The numeric calculations of real miniature bolometers show that the practical limitations are due to the thermal noise. The 1/f noise becomes important only in semiconducting bolometers in small bandwidths.

12. Acknowledgment

This work was supported by the Hungarian Research Found (OTKA) under contract no.: 77843.

13. References

Almarsi, M., Xu, B. and Castrance, J. (2006). Amorphous silicon two-color microbolometer for uncooled IR detection. *IEEE Sensor J.* vol. 6. pp. 293-300

Barocini, M., Placidi, P., Cardinali G.C. and Scorzoni, A. (2004). Thermal characterization of a microheater for micromachined gas sensor. *Sensors and Actuators A*, 115 , pp. 8-14

Bársony, I., Fürjes, P., Ádám, M., Dücső, Cs., Vízváry, Zs., J. Zettner, J., and F. Stam, F. (2004). Thermal response of microfilament heaters in gas sensing, *Sensors and Actuators B*, vol. 103 , pp. 442-447. radiation detectors

Berlicki, T.M. (2001). Thermal vacuum sensor with compensation of heat transfer. *Sensors and Actuators A*, vol. 93, pp. 27-32

Chou B.C.S., Chen Y.M., Yang, M.O. and J.S. Shie, J.S. (1996). A sensitive Pirani vacuum sensor and the electrothermal SPICE modelling. *Sensors and Actuators A*, vol. 53, pp. 273-277

Dücső, Cs., Vázsonyi, É., Ádám, M., Szabó, I., Bársony, I., Gardeniers, J. G. E., van den Berg, A. (1997).Porous silicon bulk micromachining for thermally isolated membrane formation. *Sensors and Actuators A*, vol. 60, pp. 235-239

Ebel, T., Lenggenhager R. and Baltes, H. (1992). Model of thermoelectric radiation sensors made by CMOS and micromachining. *Sensors and Actuators A*, vol. 35, pp. 101-106.

Fleetwood, D. M. & Giordano, N. (1983). Resistivity dependence of 1/f noise in metal films. *Physical Review B*, vol. 27, pp. 667-671

Fodor, Gy. (1965). *Laplace Transforms in Engineering*, Publishing House of the Hungarian Academy of Sciences, Budapest

Fürjes, P., Vízváry, Zs., Ádám, M., Morrissey, A., Dücső, Cs. and Bársony, I., (2002). Thermal investigations of a microheater for micromachined gas sensor. *Sensors and Actuators A*, vol. 115, pp 98-103

Fürjes P., Légrádi G., Dücső Cs., Aszódi A. and Bársony I. (2004). Thermal characterisation of a direction dependent flow sensor. *Sensors and Actuators A*, vol. 115, pp. 417-423

Graf, A., Arndt, M., Sauer, M. and Gerlach G. (2007). Review of micromachined thermopiles for infrared detection. *Measuring Science and Technologies*, vol. 18, pp. R59-R75.

Hennini M. & Razeghi M. (2002). *Handbook of Infra-red Detection Technologies*, Elsevier, ISBN: 978-1-85617-388-9.

van Herwaarden, A.W., & Sarro, P.M. (1986). Thermal sensors based on the Seebeck effect. *Sensors and Actuators*, vol. 10. pp. 321-346.

F.N. Hooge F.N. (1969). 1/f noise is no surface effect. *Physics Letters*, vol. 29 A, pp. 139-140

Hooge, F.N., Kleinpenning T.G.M. and Vandamme L.K.J. (1981). Experimental studies on 1/f noise. *Rep. Prog. Phys.* vol. 44, pp. 479-532.

Hooge, F.N. (2002). On the additivity of generation-recombination spectra. Part 1: Conduction band with two centres, *Physica B*, vol. 311, pp. 238-249

Hooge, F.N. (2003). On the additivity of generation-recombination spectra. Part 2: 1/f noise. *Physica B*, vol. 336, pp. 236-251.

Imran M. & Bhattacharyya A. (2005). Thermal response of an on-chip assembly of RTD heaters, sputtered sample and microthermocouples. *Sensors and Actuators A: Physical*, vol. 121. pp. 306-320

Jones, R. C. (1949). Factors of merit for radiation detectors. *J. Opt. Soc. Am.* vol. 39, p. 344-356

Jones, B.K. (1994). Low-frequency noise spectroscopy, *IEEE Trans. On El. Dev.* vol. 41, pp. 2188-2197

Kingston R.H. (Ed.), (1957). *Semiconductor Surface Physics*, University Pennsylvania Press, Philadelphia

Kingston, R.H. (1978). *Detection of Optical and Infrared Radiation*, Springer-Verlag, New York.

Kogan, Sh. (1996). *Electronic noise and fluctuations in solids*, ISBN 0 521 46034 4, University Press, Cambridge

Khrebtov, I. A., (2002). Noise properties of high temperature superconducting bolometers. *Fluctuation and Noise Letters*, vol.2. pp. R51-R70

Kruse, P.W., McGlauchlin, L.D. and McQuistan, R.B. (1962). *Elements of infrared technology.*John Willey&Sons, New York, London

Leggenhager, R., Baltes, H., Peer, J. and Forster M. (1992). Thermoelectric Infrared Sensors by CMOS Technology. *IEEE Electron Device Letters*, vol. 13. pp. 454- 456.

Liddiard, K. C. (1984). Thin-film resistance bolometer IR detectors. *Infrared Phys.* vol. 24. pp. 57-64

Liddiard, K. C. (1986). Thin-film resistance bolometer IR detectors II. *Infrared Phys.* vol. 26. pp. 43-49

Liddiard, K. C. (1993). Application of interferometric enhancement to self-absorbing thin film thermal IR detectors. *Infrared Phys.* vol. 34. pp. 379-387

Lienhard IV, J.H. & Lienhard V, J.H. (2003). A *Heat Transfer Textbook*. Phlogiston Press, Cambridge, Massachusetts, U.S.A.

Maul, M.K., Strandberg M. W. P. and Kyhl, R. L. (1969). Excess noise in superconducting bolometers. Physical Review, vol. 182. pp. 522-525

McMillan, W. L. (1968). Transition temperature of srong-coupled superconductors. *Physical Review*, vol. 167, pp. 331-344

de Nivelle, M.J.M.E., Bruijn, M.P., de Vries R., Wijnbergen J. J., de Korte, P. A. J., Sánchez, S., Elwenspoed, Heidenblut, T., Schwierzi, B., Michalke, W. and Steinbeiss, E. (1997). Low noise high-Tc superconducting bolometers on silicon nitride nembranes for far-infrared detection. *Journal of Applied Physics*, vol. 82, pp. 4719-4726

Roncaglia, A., Mancarella, F. and Cardinali, G.C. (2007). CMOS-compatible fabrication ofthermopiles with high sensitivity int he 3-5 μm atmospheric window. *Sensors and Actuators* B. vol.125. pp. 214-223

Socher E., Degani, O. and Nemirovsky, Y. (1998). Optimal design and noise considerations of CMOS compatible IR thermoelectric sensors. *Sensors and Actuators*, vol. A 71, pp. 107-115.

Szentpáli, B., Ádám, M. and Mohácsy, T. (2005). Noise in piezoresistive Si pressure sensors. *Proc. of the SPIE vol. 5846*. pp. 169-179. Austin, TX.

Szentpáli, B. (2007). Noise Limitations of the Applications of Miniature Thermal Resistors. *IEEE Sensors Journal*, vol. 7, No. 9, pp. 1293-1299

Szentpáli, B., Basa, P., Fürjes, P., Battistig G., Bársony I., Károlyi, G., Berceli, T., Rymanov, V. and Stöhr, A. (2010). Thermopile antennas for detection of millimeter waves. *Applied Physics Letters* vol.: 96, 133507, [doi.: 10.1063/1.3374445].

H.A. Watson, H. A., (1969). Microwave semiconductor devices and their circuit applications. McGaw-Hill, New York, pp. 379-381

Xu, D., Xiong, B. and Wang, Y. (2010). Modeling of Front-Etched Micromachined Thermopile IR Detector by CMOS Technology. *Journal of Microelectromechanical Systems*, vol. 19, pp. 1331-1340.

van der Ziel, A., (1986) *Noise in Solid State Devices and Circuits*, Wiley-Interscience ISBN 0-471-83234-0

Part 3

Advances and Trends

6

Bolometers for Fusion Plasma Diagnostics

Kumudni Tahiliani and Ratneshwar Jha
Institute for Plasma Research,
India

1. Introduction

The thermonuclear fusion is one of the most seriously pursued alternative sources of energy for the future of mankind. The fusion energy is safer and cleaner compared to fission energy, produces no greenhouse gases and, the nuclear fuels are evenly distributed throughout the globe. Nuclear fusion is responsible for heat and radiation generated by the Sun. In the Sun, two atoms of hydrogen fuse together to produce helium. It has been determined that the fusion of the two hydrogen isotopes, namely deuterium (D) and tritium (T) that produces 17.6 MeV (mega electron-volt) of fusion energy, is feasible in a laboratory setting (Wesson, 2004). The D-T fusion however takes place at fairly high temperature of 10-30 keV or, $(1-3) \times 10^7$ K, which is necessary for deuterium and tritium nuclei to come close together to overcome electrostatic repulsion. At these thermonuclear temperatures, the atoms get stripped of all the electrons and form a plasma (electrically charged gas). Such plasmas can be confined in a desired region by using strong magnetic fields. The magnetic fields force the particles to spiral along the field lines thus confining them. The most promising magnetic confinement systems are toroidal in shape. Among the toroidal shaped plasma devices, tokamak is the most advanced one. Presently, ITER (International Thermonuclear Experimental Reactor) is the largest tokamak under construction at Cadarache, France, and JET (Joint European Torus) in Culham, UK is the largest operating tokamak. Other non-magnetic confinement systems are also being investigated. For example, the laser induced inertial confinement systems.

One of the main requirements of the fusion plasmas is to heat the plasma particles to very high temperatures. The typical power required to attain the thermonuclear temperatures in ITER is approximately 50 megawatt (Shimada et al., 2007). The heating methods employed to heat the plasma to these temperatures are ohmic heating, neutral beam heating, and radiofrequency (RF) heating. In order to ohmically heat the plasma, a current of the order of millions of amperes is induced in the plasma. The current heats the plasma through the acceleration of charged particles and provides few megawatts of power. In the neutral beam heating method, a beam of energetic neutral particles of the working gas heats the plasma particles by momentum transfer. This provides power of the order of tens of megawatts. Radio frequency heating involves injection of RF power, either matching with the ion cyclotron frequency or the electron cyclotron frequency. This method also provides heating power of the order of 10 megawatts. Apart from temperature, there are other conditions on the density and the confinement time of fusion plasma, which comes from the breakeven criterion for the feasibility of nuclear fusion for energy production. The required conditions are ion density of $1-2 \times 10^{20}$ m^{-3} and confinement time of 4-6 seconds. The confinement time depends on several loss mechanisms through which the power is lost from the plasma. For example, plasma energy is lost through fast neutrals resulting from the charge exchange processes of plasma ions that cannot be confined by the magnetic field, through the cross-field diffusion of charged particles, and through radiation from the impurities present in the plasma. The radiation power loss is a significant fraction of the total power loss and reaches 100% in certain plasma regimes.The impurities that contribute to the radiation power loss from the plasma are low z (hydrogen, carbon, and oxygen), medium z (iron, molybdenum and silicon), and high z (tungsten). The low z impurities enter the plasma by low energy detachment processes of molecules adsorbed on the machine walls. Whereas, sputtering, arcing and evaporation processes from the wall or plasma facing components release medium and high z impurities. If the concentrations are large enough, these impurities may radiate enough power through spectral lines of incompletely stripped ions to degrade the plasma energy confinement. Ignition of D-T plasmas can be prevented by the presence of only 3% of low z-elements (oxygen), 1% of intermediate z element (iron), or 0.1% of high z element (tungsten) (Jensen et al., 1997). So it is essential to measure and control the radiation power loss in fusion plasmas. The radiated power measurement is also needed in power balance calculations, in the study of ion and electron transport processes, and also to study the interaction of plasma with walls and limiter. Some physical phenomena, for example, plasma disruptions, radiative instabilities, and detached plasmas are studied using the power loss measurements.

The radiation power loss from tokamak plasma is measured using bolometers. Bolometers are detectors that can measure radiation over a broad spectrum, from the soft x-rays to the infrared, with a nearly uniform responsivity at all wavelengths. The bolometers are used in two types of configurations in tokamaks. In the first type, a single bolometer is placed behind a pinhole and it views the whole poloidal section of the plasma from one toroidal location. This gives the total power radiated from the plasma since the tokamak plasmas are toroidally symmetric. In the second configuration, a linear array of bolometers is placed behind a single pinhole, each bolometer looking at different region of the plasma. These measurements are inverted to obtain the radiation emission distribution in the plasma.

The operational requirements of bolometers in tokamak plasmas are high sensitivity, small area for optimal spatial resolution (few mm in edge plasmas), in-situ calibration, fast time response (less than few ms), ultra-high vacuum compatibility, and high electromagnetic fields and high temperature compatibility. The tokamaks are equipped with huge systems like magnets both for confinement and for equilibrium, limiters, diverters, heating systems, and gas puff systems. These systems are spread on and around the tokamak. This renders a very limited space for diagnostics and a limited access to the ports. Difficulty to access the bolometers after mounted inside requires bolometers to be reliable over a long period of operation. The bolometers should also meet radiation hardness requirement to prevent damage by neutron and gamma radiation in present and next generation fusion devices.

Different kinds of bolometers have been tested in tokamaks for radiation power loss measurements. Bolometers used in currently operating fusion plasmas have evolved from the ones used in other non-fusion plasmas. These bolometers need some changes to adapt in future fusion devices. Following section briefly discusses all kinds of radiation sensors/detectors that have been used in the plasma devices for radiated power measurements. It describes in detail the detectors that are used in currently operating devices. The section also evaluates feasibility of such detectors for use in next generation fusion devices.

2. Types of bolometers

In the earlier tokamaks, the radiation was measured using thermal detectors. A thermal detector absorbs radiation that in turn changes its physical property. It is also sensitive to the particle flux. So, in addition to the radiation, thermal detectors also measure the energy loss from neutral and charged particles. Thermopiles and pyroelectric detectors are two types of thermal detectors that have been used in tokamaks. Bolometer refers to the kind of thermal detector in which the temperature change produced by the absorption of radiation and particle flux causes a change in the electrical resistance of the detector material. Thermistor, semiconductor, and metal foil are the types of bolometers that have been used in tokamaks. AXUV (Absolute Extreme Ultra Violet) detectors and IR (Infra Red) detectors are fairly new techniques for measuring the radiated power. AXUV detectors are photodiodes that measure the current induced by the incident radiation. Whereas, IR detectors measure the IR radiation emitted by the metal foil placed in the line of sight of the incident radiation.

2.1 Thermopiles

Thermopiles have several thermocouples, connected in series. Its principle of operation is same as of a thermocouple. The most commonly used materials for a thermopile are antimony and bismuth, which give the best seebeck coefficient. The responsivity of a thermopile is given by,

$$R_{res} = \frac{n\alpha\varepsilon}{\kappa} \tag{1}$$

where, n is the number of thermocouples, α is the seebeck coefficient, ε is the detector emissivity and κ is the thermal conductance. The typical responsivity is 5-15 V/W and the time response is few milliseconds (Sharp et al., 1974). They have been used in ORMAK (Edmunds and England, 1978) and DITE tokamaks. Although the responsivity of thermopiles is high, it has a slow response time and very sensitive to radiation damage. Therefore, thermopiles are not suitable for use in fusion plasmas.

2.2 Pyroelectric detector

Pyroelectric detectors are made of ferroelectric single crystals that have permanent electrical polarization. The temperature of the material affects the degree of this polarization, and a change in the surface charge results from a change in temperature, which can be measured through an external circuit. Lithium niobate or tantalate ($LiNbO_3$ or $LiTaO_3$), deuterated triglycine sulfate (DTGS), strontium barium niobate, and polyvinylidene fluoride are some of ferroelectric materials. The current induced is given by,

$$I = \rho(T) A \frac{dT}{dt} \tag{2}$$

where, $\rho(T)$ is the pyroelectric coefficient, A is the sensing area, and dT/dt is the rate of temperature change. Pyroelectric detectors can directly measure differential radiation power. But there is a need for a preamplifier near the detector. Also it is sensitive to radiation damage, which begins to take effect from a 1 MeV neutron fluence of $\sim 5 \times 10^{14}$ n-cm^2 (Orlinskiz and Magyar, 1988). These detectors have been used in DIVA (Odajima et al.1978), ISX-B (Bush and Lyon, 1977) and TFR (TFR groups, 1980) tokamaks.

2.3 Thermistor

Thermistors are made of materials that have a high temperature coefficient of resistance. They include germanium and oxides of manganese, cobalt, or nickel. The absorbing element is made by sintering wafers of these materials together and mounted on an electrically insulating but thermally conducting material such as sapphire. The incident radiation increases the temperature of the thermistor, hence, decreasing the resistance. Thermistors are highly sensitive and have been used in Alcator A (Scaturro and Pickrell, 1980), PBX (Paul et al., 1987), and ATC (Hsuan et al., 1975) tokamaks. But they are damaged by even a relatively small dose of neutrons and gamma rays ($\sim 10^7$ rad) (Schivell, 1982).

2.4 Semiconductor foil bolometer

Semiconductor foil bolometers consist of a metallic absorber foil and a semiconductor detector film connected together by a thermally conducting and electrically insulating substrate. Under normal operation, an accurately controlled bias current is passed through the resistor element. Thus for a radiation input causing a change in the bolometer resistance, there is an output voltage,

$$\Delta V = IR_0(1 - \exp(-\alpha \Delta E / C)) \tag{3}$$

Here, R_0 is semiconductor resistance at 0 °C, α is temperature coefficient of resistance (4-5 % K^{-1}), ΔE is absorbed energy, and C is heat capacity of the multilayer foil. These bolometers have been used in TM-2, TM-3 (Gorelik et al., 1972), JFT-2 (Maeno and Katagiri, 1980), and TCA (Joye and Marmillod, 1986) tokamaks. Despite their high sensitivity, these bolometers are prone to radiation damage.

2.5 Metal foil bolometer

A typical metal foil bolometer is composed of three layers: a metallic absorber layer, a thermally conducting but electrically insulating substrate layer, and a metallic resistor layer (see figure (1)). The absorber layer absorbs the incident radiation, which causes a rise in its temperature. The heat is conducted through the substrate layer to the resistor layer, which increases its temperature and hence the resistance. The change in resistance is measured electrically and is related to the incident power.

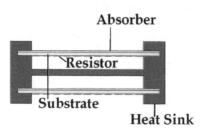

Fig. 1. Schematic of metal foil bolometer configuration, showing the measuring and the reference bolometer.

The absorber layer is made of metal, which is either gold or platinum. These metals have nearly constant absorption from 2000Å up to soft x-rays and a high reflectivity for wavelengths below 2000Å (Sabine, 1939). The substrate layer made of either mica or kapton provides mechanical strength to the bolometer and electrically insulates the absorber layer from the resistor layer. The resistor layer is of the same metal as the absorber layer and is made in the form of a meander to have a high resistance (~ few kΩ). The whole foil is in contact with a lateral heat sink.

The thickness of the metal foil bolometer is a compromise between two opposing requirements: (1) the absorber layer has to be thick enough to absorb the highest energy radiation expected from the plasma, and (2) the total thickness of the bolometer has to be small in order to increase sensitivity and decrease response time. The typical thickness of the absorber layer is 3-4 μm, of the substrate layer is around 7 μm, and that of the resistor layer is around 0.01μm.

To compensate for temperature drifts and electromagnetic interferences, a second reference bolometer, shielded from incident plasma radiation, is used. A reference meander and another measurement resistor are coupled in a bridge circuit such that the output voltage is proportional to the temperature change of the measuring bolometer.

2.5.1 Metal foil bolometer signal analysis

If a radiation power P is uniformly incident on bolometer foil, the temporal evolution of the temperature is governed by the differential equation (Mast et al., 1991),

$$P = C\left(\frac{dT}{dt} + \frac{T}{\tau_c}\right) \tag{4}$$

Here, C (= A $c_p\rho d$) is heat capacity of the foil and τ_c is the cooling time constant, which is a measure of heat loss rate to the heat sink. The radiation losses have been omitted since the detector temperature is close to the ambient temperature for most of the cases. Also, a uniform profile has been assumed on the foil.

The frequency response of the bolometer is given by,

$$T(\omega) = \frac{P(\omega)}{C}\frac{\tau_c}{\sqrt{1+\omega^2\tau_c^2}} \tag{5}$$

Thus, at a fixed incident power $P(\omega)$, the temperature is inversely proportional to the heat capacity. So the bolometer thickness should be minimum for a higher signal level. Also, when the frequency is low i.e. $\omega\tau_c \ll 1$, the temperature amplitude is given by,

$$T(\omega) = \frac{P(\omega)}{C}\tau_c \tag{6}$$

Thus, the temperature grows linearly with τ_c at low frequencies. At high frequencies of the input power ($\omega\tau_c \gg 1$)

$$T(\omega) = \frac{P(\omega)}{C\omega} \tag{7}$$

In this case, the temperature is inversely proportional to the input frequency and the thermal insulation of the bolometer foil has no effect on the signal to noise ratio.

In equation (4), a uniform incident power as well as uniform temperature distribution on the foil was assumed, which is not true in the actual case. The heat conduction to the sink results in losses and a non-uniform temperature distribution. This heat conduction is in X-Y-Z direction but since the thickness of the bolometer is very small compared to the area, a two-dimensional treatment of heat conduction is sufficient. The heat diffusion equation for the bolometer can be written as (Mast et al., 1991),

$$P(x,y,t) = C\left[\frac{\partial T(x,y,t)}{\partial t} - D\left(\frac{\partial^2}{\partial x^2} + \frac{\partial^2}{\partial y^2}\right)T(x,y,t)\right] \tag{8}$$

Here D, is the thermal diffusivity. For a square foil of side 2a, T can be expanded in terms of double Fourier series, with mode numbers (l, m).

$$T(x,y,t) = \sum_l \sum_m T_{l,m}(t)\cos\left[(2l+1)\frac{\pi x}{2a}\right] \times \cos\left[(2m+1)\frac{\pi y}{2a}\right] \tag{9}$$

Similarly, we can write P as,

$$P(x,y,t) = \sum_l \sum_m P_{l,m}(t)\cos\left[(2l+1)\frac{\pi x}{2a}\right] \times \cos\left[(2m+1)\frac{\pi y}{2a}\right] \tag{10}$$

Using above equations, we get the following solution for equation (8),

$$P_{l,m}(t) = C\left[\frac{d}{dt}T_{l,m}(t) + \frac{T_{l,m}(t)}{\tau_{l,m}}\right] \tag{11}$$

where,

$$\tau_{l,m} = \frac{\tau_{00}}{2\left((2l+1)^2 + (2m+1)^2\right)} \tag{12}$$

Time constant τ_{00} corresponds to the fundamental mode of the series expansion. The fundamental Fourier mode (0,0) contributes maximum to the solution with rest of the modes up to (2,2) making some contribution (up to 3%) to the sum. Considering only the fundamental mode and assuming constant radiation power, the solution is same as equation (4).

Using equation (4), the temperature of the foil can be written as,

$$T = \frac{P\tau_c}{C}\left(1 - e^{-t/\tau_c}\right) \tag{13}$$

The temperature increase causes an increase in the resistance of the metal given by,

$$\Delta R = R_0 \alpha \Delta T \tag{14}$$

which is measured using a standard Wheatstone bridge (Schivell et al., 1982).

2.5.2 Calibration procedures

A well developed method is required to calibrate the detectors in order to determine their physical properties of use. The aim of the calibration procedures in the metal foil bolometers is to determine the value of the heat capacity C and the cooling time constant τ_c. This is done in two stages. First, a known constant power P is made incident on the detector, which raises the temperature of the foil to a saturation value, T_{sat}. The power and the T_{sat} measurements give the value of ratio τ_c / C, which is also called the thermal resistance (z) of the bolometer,

$$z = \frac{\tau_C}{C} = \frac{T_{sat}}{P} \tag{15}$$

In the second stage, the bolometer surface is uniformly irradiated by using a radiation pulse and then the radiation power source is turned off. This results in the exponential decay of the temperature, known as the cooling curve. From the cooling curve of the bolometer, τ_C is determined using Eq. (13). Using the measured values of τ_C and z, the value of C is determined.

The metal foil bolometer is presently the most widely used detector in tokamak plasmas. It has been used in TFTR (Schivell et al., 1982), ASDEX (Muller and Mast, 1984), JET (Mast et al., 1985) and TEXT (Snipes et al., 1984) tokamaks. Table 1 lists the metal foil bolometer characteristics for JET and ASDEX tokamaks. Metal foil bolometer has been used in D-T plasmas and has shown resistance to radiation up to 10 Grad. The metal foil bolometer is the potential candidate for ITER tokamak, and it is being tested for the expected radiation and neutron flux levels.

Bolometer foil characteristics	JET (Mast et al., 1985)	ASDEX (Muller and Mast, 1984)
Heat capacity C (mJ K^{-1})	2.2	2
Cooling time constant τ_C (s)	0.2	0.173
Heat resistance z = τ_C / C (K/W)	90	90
Detection limit (μW/cm^{-2})	70	100
Response time (ms)	1	1
Resistance R (kΩ)	4.8	5
Resistance temperature coefficient α ($\Delta R/\Delta T \times 1/R$)(K^{-1})	2.7 x 10^{-3}	2.5 x 10^{-3}
Response dV/dP (I = 1mA) (V/W)	0.5	1.2
Integration time τ_{int} (ms)	20	10
Foil size	11 x 11 (mm^2)	10 x 10 (mm^2)
Baking temperature ($^\circ$C)	150	150
Neutron fluence cm^{-2}	2.5 x 10^{14}	2.5 x 10^{14}
Radiation dose (rad)	1 x 10^{10}	1 x 10^{10}
Foil/substrate	Gold/Kapton	Gold/Kapton

Table 1. Metal foil bolometer parameters in JET and ASDEX.

2.6 AXUV detectors

AXUV detectors are special kind of photodiodes that have a uniform response from the ultraviolet up to the soft x-ray energies. Their basic operation is similar to a photodiode. When a photon of energy hv > E_g (band gap energy) is incident on a photodiode, electron-hole pairs are produced and are swept apart by the internal electric field at the p-n junction. This generates a current, which is called photocurrent. The upper wavelength detection limit is determined by the attenuation coefficient of the material of the photodiode and the thickness while the lower limit is governed by the reflection and the absorption in the oxide layer at the top.

AXUV detectors are n- on p- type photodiodes with a thin silicon dioxide (SiO_2) layer and a fully depleted active region. This configuration has an advantage over the p- on n-configuration. Radiation also generates electron-hole pairs in the oxide layer. These holes accumulate close to the SiO_2/Si interface, leading to a positive charge of the oxide. This charge attracts electrons created in the bulk towards the p-doped layer adjoining the SiO_2/Si interface, where the recombination probability is higher. This results in lower responsivity. To avoid this loss, the diode structure is inverted (n on p). The positive oxide charge created in irradiation repels the electrons from the SiO_2/Si interface, thus supporting the charge carrier drift in the n-p junction when the oxide is attached to the n doped layer.

The cross section of AXUV photodiode is shown in figure (2) (Korde, 2007). The top-most layer of the device is an active oxide region (SiO_2), which acts both as an antireflection coating and a passivation layer to protect the diode. This layer is made very thin (3-7 nm) in order to reduce losses at lower wavelengths. This is very crucial since below 700 eV, a great part of the incident radiation is absorbed in the first several hundred nanometers of a semiconductor detector (Krumery and Tegeler, 1990). If a photon is absorbed in the oxide layer, possibly there is no contribution of the charge generated to the external current due to the absence of an electric field. A thin oxide layer minimizes these losses. The oxide layer in AXUV detector is also nitrided for radiation hardness of 1 Grad.

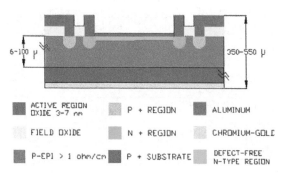

Fig. 2. The cross-section of AXUV photdiode.

2.6.1 AXUV photodiode characteristics

A photodiode performance is characterized by two parameters: the quantum efficiency (η) and the responsivity (R_{res}). Quantum efficiency is the number of electron hole pairs generated per incident photon. It is given by,

$$\eta \equiv \frac{E_{ph}}{E_g} \qquad (16)$$

where, E_{ph} is the energy of the incident photon and E_g is the band gap energy. Since the band gap energy (also called mean electron hole pair energy) is independent of the energy of the absorbed radiation, the quantum efficiency of an ideal semiconductor should linearly increase with the photon energy. A typical quantum efficiency plot of AXUV diode is shown in Figure 3. As shown, η is linear for energies above 100 eV (Check this number, as SiO2 band gap is below 20). This value can be derived theoretically from equation (16) using $E_g = 3.66$ eV for silicon (although the band gap energy is 1.1 eV in silicon, since it is an indirect semiconductor, an electron in the valence band goes to conduction band only if a phonon participates in the process. Hence the energy needed for creation of one electron hole pair is 3.66 eV). Below 100 eV, there is a significant loss in the quantum efficiency due to absorption of radiation in the oxide layer and reflection from the surface.

The second and important characteristic of a photodiode is its responsivity. It is defined as the current produced by the photodiode per unit incident power, and s given by:

$$R_{res} = \frac{I}{P} \qquad (17)$$

Here, I is the photocurrent produced by incident power, P. The quantum efficiency and responsivity are related to each other as,

$$R_{res} = \frac{\eta e}{h\nu} \qquad (18)$$

The responsivity curve of the AXUV photodiode is also shown in figure 3. The diode has a high reponsivity (0.27 A/W) at high energies and an average responsivity of 0.24 A/W above 100 eV. Below 100 eV, R_{res} varies significantly and drops down to 0.12 A/W at 10 eV.

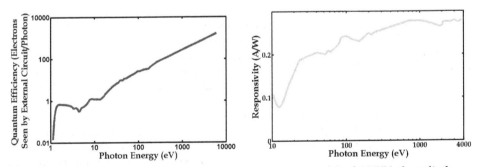

Fig. 3. Quantum efficiency (left) and the responsivity curve (right) of AXUV photodiode (Korde, 2007).

Because the response of the AXUV photodiode varies significantly at low energies, for the estimation of total radiated power from a tokamak, an effective responsivity value is determined. For this purpose the power contribution of different wavelengths to the measured is determined using different filters (Gray et al., 2004).

The response time of an AXUV bolometer is of the order of a fraction of μs. The AXUV photodiodes are ultra high vacuum compatible and can survive high baking temperatures without any significant change of properties. They can be miniaturized and made into arrays. They are also radiation hard and can survive up to 1 Grad. AXUV photodiodes have been used in TEXT-U (Wen and Brevenec, 1995), TCV (Ferno et al., 1999), and DIII-D tokamaks.

2.7 IR video bolometer

An alternative method of measuring temperature of the metal foil is to measure the infrared (IR) radiation emitted by it. The IR radiations emitted by a metal is a function of its temperature and changes as the temperature change. The IR radiations are measured using an IR camera. A prototype IR video bolometer has been successfully used in Jt-60U (Peterson et al., 2008). It has all the advantages of the metal foil bolometer, but without any electrical connection. The disadvantages of IR bolometers are slow response time and significant involvement of optics.

3. Review of bolometer diagnostic measurements

Bolometry is one of the basic diagnostics in all fusion devices. The radiation power loss measured using bolometers is crucial to the understanding of power balance of the plasma and its confinement. It also is an indication of the purity of the plasma. The radiation emission distribution and its time evolution obtained from arrays of bolometers provide information about the nature of the impurities and their transport. They can also be used to determine the nature of specific modes of operation, namely, radiative mode, detached plasmas and Marfes (multifaceted asymmetric radiation from the edge). A review of the studies carried out with bolometers is given in the following sub-sections.

3.1 Power balance of the plasma

The input power into a fusion device like tokamak can be accounted as the sum of the stored power in the hot plasma, the power conducted to the limiter/divertor or the plasma facing components and the power radiated isotropically by the hot plasma. The radiated power loss in most of the devices has been found to be a significant fraction of the input power and equals the input power in some cases. This fraction depends on the plasma parameters and also on the material of the vessel in which the plasma is contained. The parametric dependence of radiation power loss has been studied in various devices. Although no universal scaling laws have been established, general trends have been seen in many tokamaks. The radiation power loss is directly related to the plasma electron density and inversely to the toroidal magnetic field. It also has a strong dependence on the vessel conditioning and vessel material. Also, the radiation power loss data from different machines indicate variation with the size of the plasma. The larger the plasma size, the higher is the radiation power loss from it.

3.2 Radiation emission distribution

The emission of radiation is non-uniform in the plasma. The plasmas with similar total radiated power and similar conditions behave differently when the emission distributions are different. This spatial variation in emission results from the difference in the electron temperature profile and in the type of the impurity present in the plasma. The emission distribution is obtained by using two or three arrays of bolometers that cover the entire plasma cross-section. The bolometers in the array measure line integrated radiation along different lines of sight. These measurements are then inverted using a suitable inversion algorithm to determine local emission value. In case a single array of bolometers is used, one has to make an assumption of circularly symmetric distribution. The emission distribution combined with the spectroscopic measurements give a complete picture of the impurity distribution and the behavior of the plasma.

The plasmas that have a high concentration of low-z elements (carbon, oxygen) radiate strongly at the plasma boundary. The strong emission cools the edge and hence the temperature is low near the walls. This reduces the sputtering and arcing from the wall and hence the metal concentration in the plasma. So the temperature is high in the center and the plasma channel is narrow. Such discharges have good confinement. In contrast, only high-z impurities can reach the plasma core and when the concentration of high-z impurities is significantly high, it gives rise to core cooling. As a result the temperature profile is hollow, double-tearing plasma instability arises and plasma confinement is poor.

The anomaly in the radiation profile has been seen as a precursor to the density limit disruptions in some tokamaks. In both ohmic and auxiliary heated discharges, the plasma density can only be increased up to a point when the radiated power equals the input power. As the density is increased, the radiated power also increases and hence the plasma channel shrinks. This can give rise to thermal/radiative instability and Marfe (multifaceted asymmetric radiation from edge). They have been observed in several tokamaks, for example, Textor (Rapp et al., 1999), ASDEX (Stabler et al., 1992), and JET (Behringer et al., 1986).

3.3 Physics of Marfe and detached plasma

Marfe is a radiation instability that occurs when the density of a discharge is increased in plasmas with low impurity content. The increase in the density increases the radiation power ($P = L_z (T_e) n_z n_e$), which causes edge cooling and further increase of radiation power loss. The empirical scaling shows that there is a limit to the maximum achievable density at given plasma current, which is known as the Greenwald limit, n_{Gw} (m^{-3}) = 1 x 10 ^{17} I_p / πa^2 (kA/m^2). When the density of the plasma is raised towards this limit, a Marfe appears. Large amount of radiation is emitted from a small region on the high field (inboard) side of the plasma edge during a Marfe. Marfe has a certain poloidal extent (~30°) above the midplane and is present at all toroidal locations. An array of bolometers that views the plasma from the top shows asymmetry in the radiation. The exact location and extent of marfe is obtained by inverting the bolometer data from bolometer arrays that view the plasma from top and side. The radiation profile gives an unambiguous picture of the Marfe location and the time profile gives its evolution. Marfes have been studied extensively in Alcator C (Lipschultz et al., 1984), JT-60 (Nishitani et al., 1990) and TFTR (Bush et al., 1988).

Detached plasmas result from a drop in the edge electron temperature below the threshold for ionization of all the plasma species. Hence, cold neutrals surround the plasma and the radiating layer is shifted to a smaller minor radius. In detached plasmas, most of the input power is radiated from a shell at the boundary of the plasma and the power and particle fluxes to the walls are greatly reduced. Detached plasmas can be used to validate the accuracy of the bolometric diagnostics on a tokamak since the only loss channel is the radiation power loss. Another bolometric signature of detached plasma is a very steep rise in the radiated power at the plasma edge. The detached plasma regimes have been studied extensively in TFTR (Bush et al., 1988), Textor (Tokar, 1995) tokamaks, and they can be induced in order to reduce the power load on the plasma facing components.

3.4 Impurity injection for reducing heat load on the limiter and divertor

The high temperature in the thermonuclear fusion also raises the question of power exhaust at the divertor. The conducted heat loads on the limiters/divertors of the present operating devices are enormous and the ones that are expected in the next generation devices are much higher ($\sim$20 MWm^{-2}) that are impractical for any material. Thus, it is essential to device a method to reduce the heat loads. In many tokamaks, this is done by injecting trace impurities ($n_z/n_e \sim 10^{-3}$) of gases with $z \geq 10$ in the periphery of the plasma without affecting the core plasma parameters and stability. In TEXTOR (Samm et. al., 1993) and Tore Supra (Grosman, 1995), it has been demonstrated that the injection of neon enabled establishment of quasi-stationary layer that radiated 90% of the input power (Pospieszczyk et. al., 1995) from the edge. In TFTR neutral beam heated discharges trace impurity of xenon, and krypton were used to generate discharges with increased radiation power loss and decreased conduction and convection power losses to the limiter (Hill et al., 1999).

4. Examples from Aditya tokamak

We now present examples of radiation power measurements on Aditya tokamak. It is a medium sized tokamak of major radius 0.75m and minor radius 0.25m (Bhatt et al., 1989). It is in operation for two decades, and during this period various experiments have been conducted to study the edge fluctuations, turbulence and plasma instabilities. The typical operational parameters of Aditya plasmas are as follows: toroidal magnetic field on axis $B_T = 0.75$ T, plasma current $I_p = 75\text{-}100$ kA, central electron density $n_e = (1\text{-}3) \times 10^{19}$ m^{-3} and central electron temperature $T_{e0} = 300\text{-}400$ eV. It is equipped with all the standard diagnostics.

4.1 Bolometer system

There are two cameras for the measurement of the radiation power loss and radiation emission distribution in Aditya. One is a single channel collimated bolometer (AXUV-5), mounted on the top port that views the whole poloidal plasma cross-section (solid angle 0.16 sr). A second camera that has an array of 16 detectors (AXUV-16ELO) is mounted on the radial port and views the whole poloidal cross section of the plasma through a pinhole. The spatial resolution of this camera is 10 cm at the vessel mid-plane and the temporal resolution is 0.2 ms. The detector array and the single detector are mounted in UHV chambers and connected to the electronics in a compact housing mounted on the machine port. The electronics includes current-to-voltage convertors, amplifiers, filters and drivers for all photodiodes. The fully differential output from the electronics card is then transmitted to the data acquisition system (Tahiliani et al., 2009).

4.2 Bolometer data analysis

The AXUV detector signal (S) is related to the radiation emission in the plasma and can be written as,

$$S = G \, R_{res} \, E \int g(r) \, dl \qquad (19)$$

Here, G (V/A) is amplifier gain, R_{res} (A/W) is AXUV responsivity, E (m²) is etendue of the system , and g (Wm-3) is the emissivity, which is a function of the electron density, impurity density, and the electron temperature. The integral is along the line of sight of the detector. The line-integrated emissivity, is also called the brightness, B.

The radiation power loss P_{rad} (W) can be obtained from the measured value of B. For the single channel detector P_{rad} is given by,

$$P_{rad} = 2\pi R_0 \, 2a \, B \qquad (20)$$

It can also be determined from the array measurements by the following equation:

$$P_{rad} = 2\pi R_0 \sum_i B_i \Delta r_i \qquad (21)$$

Here R_0 is major radius of the tokamak, a is minor radius, Δr_i is chord width at the mid-plane, and the summation is over all detectors of the array.

The line-integrated measurements B of the array detectors can be inverted using a suitable inversion algorithm to obtain emissivity (g) distribution in the plasma. In Aditya, ART (Algebraic Reconstruction Technique) is used (Tahiliani et al., 2009) along with an assumption of circular symmetry.

4.3 Experimental results

An example of measured B-profile is shown in figure 4. It is observed that radiation power loss from Aditya tokamak ranges from 20% - 40% of the input power during the current flat top and is close to the input power at the end of the discharge. It has been shown separately that the radiated power fraction (P_{rad} /P_{in}) decreases linearly with increasing current indicating that the low-z impurities are dominant since they radiate less at elevated temperatures. The variation of radiation power loss has also been studied with the electron density. It is seen that the radiated power remains constant with increasing density up to $[n_e] \sim 1.6 \times 10^{19}$ m-3 and thereafter it increases with the density.

The radiation emission profiles are found to be hollow with little radiation from the central part of the plasma column and a radiation peak at the edge. This is in line with the presence of low-z impurities and a very low metal concentration. With the central electron temperature of 300 – 400 eV in Aditya we expect the low-z impurities to be fully ionized.

We have further carried out radiation power measurements in discharges with additional short pulses of the working gas (Gas Puff, GP) as well as MBI (Molecular Beam Injection). It is observed that GP leads to an increase in the edge radiation while MBI reduces the recycling in the edge. This may indicate that GP may be used to cause radiation cooling of the plasma edge and thereby reduce plasma instability drive that is responsible for edge fluctuation (Jha et al., 2009). On the other hand, MBI can also be used for edge modification, probably by causing velocity shear suppression of fluctuation because both seem to reduce edge recycling.

Radiation power measurements in discharges with DLD (density limit disruptions) have also been studied. It is observed that the edge detector signal increases during the disruption phase and becomes comparable to that of the central detector indicating contraction of the plasma channel. In the DLD induced by GPs we observe that the edge line-integrated radiation is about an order of magnitude higher than the core line-integrated radiation. However, because of the fact that during DLD the edge detectors saturate and there may be significant radiation loss in the wavelength range 1200 – 1500 Å, in which detector sensitivity is poor, it is not possible to estimate the radiation fraction accurately.

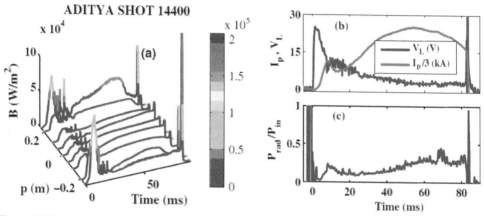

Fig. 4. (a) The brightness (B) profile of the array detectors for Aditya discharge number 14400, (b) plasma current (I_p) and loop voltage (V_L) that gives input power $P_{in}=I_pV_L$, and (c) the fraction of the radiated power (P_{rad}) to the input power P_{in}.

5. Bolometers on ITER tokamak

ITER is an international collaboration to test the feasibility of fusion and it is under construction in Cadarache, France. ITER aims to demonstrate that it is possible to produce commercial energy from fusion. The scientific goal of the ITER project is to deliver ten times the power that it consumes. From 50 MW of input power, the ITER machine is designed to produce 500 MW of fusion power – the first of all fusion experiments to produce net energy. During its operational lifetime, ITER will test key technologies necessary for the next step: the demonstration fusion power plant that will prove that it is possible to capture fusion energy for commercial use. The physics design of ITER is completed (Perkins et al., 1999) and its construction is underway.

In ITER, the use of bolometer diagnostic is two folds. The first is to measure and control the radiation power loss from the divertor region. For the sustenance of the preferred operation at high density, it is critical to control and enhance the power that is radiated from the edge of the plasma, especially from the divertor legs and the X-point. The second is to measure the spatially resolved radiation power for the study of the power balance of the discharge. A spatial resolution of 20 cm in the main plasma and 5 cm in the divertor and X-point region is required for these tasks. Also, the proposed method for the inversion of the bolometer data is sparse data tomography, which would require 340 lines of sights of measurements (Young et al., 1999). The bolometer arrays will be installed in the equatorial and vertical ports and in the specially instrumented divertor diagnostic cassettes. From each of these locations, several arrays of detectors will observe the plasma. From the equatorial port, the inner divertor leg (high resolution) and the main plasma are viewed; from the vertical port, the main plasma, the area of the X-point, and the largest part of the divertor legs (high resolution) can be seen. This view can give the total radiation power loss.

Efforts are being made to develop and test a suitable bolometer detector that could withstand the high neutron flux and the temperature expected in ITER. The typical values of the expected neutron flux and radiation dose (mostly gamma ray) are listed in table 2. It is seen that neutral flux is five times, neutron flux is ten times and neutron fluence is 10,000 higher compared to the existing machines.

Parameters	First wall	Blanket gap (VV)
Neutron flux $[m^{-2}s^{-1}]$ (>0.1 MeV) (14 MeV)	3×10^{18} 8×10^{17}	$(0.2\text{-}1) \times 10^{17}$ $(0.8\text{-}4) \times 10^{16}$
Fluence (>0.1 MeV) $[m^{-2}]$	3×1025	$(0.4\text{-}2) \times 1024$
Dose rate $[Gy/s = 100\ rad/s]$	2000	20-100
Neutral particle flux $[m^{-2}s^{-1}]$	5x1019	1x1018
Plasma radiation $[kW\ m^{-2}]$	500	10
Time varying mag. field [T]	1	~1

Table 2. The estimated particle and radiation load in ITER.

The most eligible candidate for ITER is the metal foil bolometer. The metal foil bolometer used on JET has a proven performance in an ITER like environment. The metal foil bolometers presently being used on JET tokamak (see table 1) have been tested and the neutron irradiation tests show that they can't survive a full lifetime of ITER due to the transmutation of gold to mercury, embrittlement of the mica or kapton foil, and the failure of the contacts. The gold has been replaced by platinum as the thermal neutron capture cross section of platinum is a factor of 10 smaller than that of gold. A new bolometer with platinum absorber (3µm) on SiN foil has been produced and tested successfully (Meister et al., 2008). But for ITER, a bolometer foil thickness of 12µm is required to ensure absorption of X-rays with energies up to 25 keV. So efforts are being done to develop a bolometer with a thickness of 12 µm.

6. Conclusion

The bolometers are essential diagnostic for the measurement and control of the radiation power loss from the fusion plasmas. This diagnostics is used extensively both for physics studies and for operation and control of fusion plasma devices. Over the last thirty years, this diagnostics have been developed from the early use of thermister and thermopiles to the present use of absolute photodiodes and radiation hard metal foil bolometers. The present day bolometers can withstand neutron fluence up to 2.5×10^{14} cm^{-2} and radiation dose up to 10 Grad. However, the next generation fusion devices will have much higher levels of radiation doses and neutron fluence, and therefore new bolometers that work in such harsh conditions may have to be developed.

7. References

Behringer, K.H., Carolan, P.G., Denne, B.,Decker, G., Engelhardt, W., Forrest, M.J.,Gill, R.,Gottardi, N. Hawkes, N.ckalline, E., Krause, H. Magyar, G. Mansfield, M.,Mast F., Morgan, P., Peacock, N.J., Stamp, M.F. & Summers H.P. (1986). Impurity and Radiation Studies During The JET Ohmic Heating Phase. *Nuclear Fusion*, Vol.26, No.6 (1986) pp.751-768, ISSN 0029-5515

Bhatt, S.B., Bora, D., Buch, B.N., Gupta, C.N., Jain, K.K., Jha, R., John, P.I., Kaw, P.K., Kumar, Ajay, Matto, S.K., Natarajan, C., Pal, R., Pathak, H.A., Prabhakara, H.R., Pujara, H.D., Rai, V.N., Rao, C.V., Rao, M.V.V., Sathyanarayan, K., Saxena, Y.C., Sethia, G.C., Vardharajalu, A., Vasu, P. & Venkataramani, N. (1989). ADITYA: The First Indian Tokamak. *Indian Journal of Pure and Applied Physics*, Vol.27, (1989), pp. 710-742, ISSN 0019-5596

Bush, C.E. & Lyon, J. F. (1977). Wall Power Measurements of Impurity Radiation in ORMAK.*Oak Ridge National Laboratory Report*, ORNL/TM-6148, (December 1977)

Bush, C.E., Schivell, J., McNeill, D.H., Medley, S.S., Hendel, H.W., Hulse, R.A., Ramsey, A.T., Stratton, B.C., Dylla, H.F., Grek,B., Johnson, D.W., Taylor, G., Ulrikson, M. & Weiland, R.M. (1988). Characteristics of Radiated Power for Various Tokamak Fusion Test Reactor Regimes. *Journal of Vacuum Science and Technology A*, Vol. 6, No.3, (May/June 1988), pp 2004-2007, ISSN 0734-2101

Edmonds, P.H. & England A.C. (1978). Eneregy Loss To The Wall and Limiter in NormalORMAK Discharges. *Nuclear Fusion*, Vol.18, No.1 (1978), pp. 23-27, ISSN 0029-5515

Furno, I., Weisen, H., Mlynar, J., Pitts, R. A., Llobet, X., Marmillod, Ph. & Pochon, G. P. (1999)Fast Bolometric Measurements on The TCV Tokamak. *Review of Scientific Instruments*, vol. 70, No.12, (December 1999), pp. 4552-4556, ISSN 0034-6748

Gorelik, L.L., Mirnov, S.V., Nikolevsky, V.G. & Sinitsyn, V.V. (1972) Radiation Power of aPlasma as a Function of the Discharge Parameters in the Tokamak-3 Device *Nuclear Fusion* Vol.12 (1972) pp. 185-189, ISSN 0029-5515

Gray, D. S., Luckhardt, S. C., Chousal, L., Gunner G., Kellman, A. G. & Whyte, D. G., (2004)Time resolved radiated power during tokamak disruptions and spectral averaging of AXUV photodiode response in DIII-D *Review of Scientific Instruments* Vol. 75, No. 2 (Feburary 2004), pp. 376-381, ISSN *0034-6748*

Grosman, A., Monier-Garbet, P., Vallet, J.C.,Beaumont, B., Chamouard, C. , Chatelier, M.,DeMichelis, C., Gauthier, E., Ghendrih, P., Guilhem, D. , Guirlet, R. , Lasalle, J. , Saoutic, B. & Valter, J. (1995) Radiative layer control experiments within an ergodized edge zone in Tore supra. Journal of Nuclear Materials, Vol. 220-222, (April 1995), pp. 188-192, ISSN 0022-3115

Hill, K. W., Scott, S.D.,Bell, M., Budny,R., Bush, C. E., Clark, R. E. H., Denne-Hinnov, B., Ernst, D. R., Hammett, G. W. , Mikkelsen, D. R. , Mueller, D. , Ongena, J. , Park, H. K., Ramsey, A. T., Synakowski, E. J. , Taylor, G. & Zarnstorff, M. C. (1999) Tests of local transport theory and reduced wall impurity influx with highly radiative plasmas in the Tokamak Fusion Test Reactor. *Physics of Plasmas*, Vol.6, (1999), pp. 877-884, ISSN 1070-664X

Hsuan, H., Bol, K. & Ellis, R.A. (1975). Measurement of the Energy Balance in ATC Tokamak.Nuclear Fusion, Vol.15 (1975) pp. 657-661, ISSN 0029-5515

Korde, R., (2007) International Radiation Detectors Inc., Available from: <http://www.ird-inc.com>

Jensen, R.V., Post, D.E., Grasberger, W.H., Tarter, C.B. & Lokke, W.A. (1977). Calculations of Impurity Radiation and its Effects on Tokamak Experiments. Nuclear Fusion, Vol. 17, No.6, (1977), pp.1187-1196, ISSN 0029-5515

Jha, R., Sen, A., Kaw, P. K., Atrey, P. K., Bhatt, S. B., Bisai, N., Tahiliani, K., Tanna, R. L. & the ADITYA Team. Investigation of gas puff induced fluctuation suppression in ADITYA tokamak. (2009). Plasma Physics and Controlled Fusion, Vol.51, No.9, (2009), pp.17, ISSN 0741-3335

Joye, B., Marmillod, Ph. & Nowak S. (1986). Multichannel Bolometer for RadiationMeasurements on the TCA Tokamak. Review of Scientific Instruments, Vol.57, pp. 2449-2456, (1986), ISSN: *0034-6748*

Krumery, M. & Tegeler, E., (1990) Semiconductor photodiodes in the VUV: Determination of layer thicknesses and design criteria for improved devices. Nuclear Instruments and Methods in Physics Research Section A: Accelerators, Spectrometers, Detectors and Assocaited Equipment, Vol. 288, No.1, (1 March 1990), pp. 114-118, ISSN 0168-9002

Lipschultz, B., LaBombard, B., Marmar, E. S., Pickrell, M. M., Terry, J. L., Watterson, R. &Wolfe, S. M. (1984) Marfe: An Edge Plasma Phenomena. Nuclear Fusion, Vol. 24, No. 8, (1984), pp. 977-988, ISSN 0029-5515

Maeno, M. & Katagiri, M. (1980). An Application of a Germanium Film Bolometer toRadiation Loss Measurement in JFT-2 Tokamak Japanese Journal of Applied Physics, Vol.19, (1980), pp. 1433-1434, ISSN 0021-4922

Mast, K. F., Krause, H., Behringer, K., Bulliard, A. & Magyar G. (1985) Bolometric diagnostics in JET Review of Scientific Instrument, Vol. 56, (1985), pp.969-972, ISSN 0034-6748

Mast, K. F., Vallet, J. C., Andelfinger, C. , Betzler, P. , Kraus, H. & Schramm G. (1991). A LowNoise Highly Integrated Bolometer Array for Absolute Measurement of VUV and Soft x –ray Radiation. Review of Scientific Instruments, Vol.62, (1991), pp. 744-751, ISSN 0034-6748

Meister, H., Giannone, L., Horton, L. D., Raupp, G. , Zeidner, W. , Grunda, G. , Kalvin, S.,Fischer, U., Serikov, A., Stickel, S. & Reichle R. (2008) The ITER bolometer diagnostic: Status and plans Review of Scientific Instruments Vol. 79, (2008), pp. 10F511-10F516, ISSN 0034-6748

Muller, E. R. & Mast F. (1984). A New Metal Resistor Bolometer for Measuring Vacuum Ultraviolet and Soft x Radiation. Journal of Applied Physics, Vol.55, (1984), pp. 2635-2642, ISSN 0003-6951

Nishitani, T., Itami, K.,Nagashima, K., ,Tsuji., S, Hosogane,N., YOSHIDA, H., Ando, T., Kubo, H. & Takeuchi, H. (1990) Radiation Losses and Global Power Balance of JT-60 Plasmas. Nuclear Fusion, Vol.30, No.6 (1990), pp. 1095-1105, ISSN 0029-5515

Odajima, K., Maeda, H., Shiho, M., Kimura, H., Yamamoto, S. , Nagami, M. , Sengoku, S., Sugie, T., Kasai, S., Azumi, M. & Shimonura Y. (1978) Radiation Loss and Power Balance in DIVA. Nuclear Fusion, Vol.18, No. 10 (1978) pp. 1337-1345, ISSN 0029-5515

Orlinskiz, D.V. & Magyar G. (1988). Plasma Diagnostics on Large Tokamaks. Nuclear Fusion, Vol.28, No.4, (1988), pp 611-697, ISSN 0029-5515

Paul, S. F., Fonck , R. J., & Schmidt, G. L. (1987). Operation of a Tangential Bolometer on the PBX Tokamak. Princeton Plasma Physics Laboratory report, PPPL-2432 (1987)

Perkins, F. W., Post, D.E., Uckan N.A., Azumi, M., Campbell, D.J. , Ivanov, N., Sautho, N.R., Wakatani, M., Nevins, W.M., Shimada M., Van Dam, J., Cordey G., Costley, A., Jacquinot, J. , Janeschitz, G., Mirnov, S., Mukhovatov, V., Porter, G. , Post, D., Putvinski, S., Shimada, M., Stambaugh, R., Wakatani, M., Wesley,J., Young, K., Aymar, R., Shimomura, Y., Boucher, D. , Costley, A., Fujisawa, N., Igitkhanov, Y., Janeschitz, G., Kukushkin, A., Mukhovatov, V., Perkins, F., Post, D., Putvinski, S., Rosen-bluth, M. & Wesley J. (1999). Chapter 1: Overview and summary. Nuclear Fusion, Vol.39, (1999), pp. 2137, ISSN 0029-5515

Peterson, B.J., Parchamy,H., Ashikawa,N., Kawashima,H., Konoshima,S., Kostryukov, A.Y.,Miroshnikov,I.V., Seo,D., & Omori T. (2008). Development of imaging bolometers for magnetic fusion reactors. Review of Scientific Instruments, Vol.79, (2008), pp. 10E301-10E306, ISSN 0034-6748

Pospieszczyk, A., Samm, U., Bertschinger, G., Bogen, P., Claassen, H. A.., Esser, G., Gerhauser, H., Hey, J. D., Hintz, E., Konen, L. , Lie, Y. T. , Rusbuldt, D. , Schorn, R. P.,Schweer, B., Tokar, M., Winter, J. , the TEXTOR Team, Durodie, F., Koch,R., Messiaen, A.M., Ongena,J., Telesca, G., Vanderplas, R.E., van Nieuwenhove.R., van Oost, G., van Wassenhove, G. & Weynants (1995) R. R. Study of the power exhaust and the role of impurities in the Torus Experiment for Technological

Oriented Research (TEXTOR) Physics of Plasmas , Vol. 2, No.6, (June 1995), *I SSN: 1070-664X*

Rapp, J., De Vries, P.C., Schuller, F.C., Tokar M.Z., Biel W., Jaspers, R., Koslowski H.R., Kramer-Flecken, A.,Kreter A., Lehnen, M., Pospieszczyk, A., Reiser D., Samm U. & Sergienko G. (1999) Density Limits in TEXTOR-94 Auxiliary Heated Discharges. Nuclear Fusion, Vol. 39, No. 6 (1999) pp.765-776, ISSN 0029-5515

Sabine, G. B. (1939). Reflectivities of Evaporated Metal Films in the Near and Far Ultraviolet. Physical Review, Vol. 55, (June 1,1939) pp. 1064-1069, ISSN 1943-2879

Samm, U., Bertschinger, G., Bogen, P., Hey, J.D. , Hintz, E. , Knen, L. , Lie, Y.T.,Pospieszczyk, A., Rusbiildt, D., Schorn, R.P., Schweer, B. Tokar', M. & Unterberg B. (1993) Radiative Edges Under Control By Impurity Fluxes Plasma *Physics and Controlled Fusion*, Vol.35 (1993) pp. B167-BI75, ISSN 0741-3335

Scatturo, L.S. & Pickrell M.M. (1980). Bolometric Measurements And The Role of Radiation inAlcator Power Balance. *Nuclear Fusion*, Vol.20, No.5 (1980) pp. 527-535, ISSN 0029-5515

Schivell, J., Renda, G., Lowrance, J. & Hsuan, H. (1982). Bolometer for Measurements onHigh-Temperature Plasmas. *Review of Scientific Instruments*, Vol.53, No.10, (Oct. 1982), pp.1527-1534, ISSN *0034-6748*

Sharp, L. E., Holmes L S, Scott P E & Aldcroft D. A. (1974). A thin film thermopile for neutral particle beam measurements. *Review of Scientific Instruments* Vol.45 (1974) pp.378-381, ISSN *0034-6748*

Shimada, M., Campbell, D.J., Mukhovatov, V., Fujiwara, M.Kirneva, N., Lackner, K., Nagami, M., Pustovitov, V.D., Uckan, N. , Wesley, J., Asakura, N., Costley, A.E., Donne, A.J.H., Doyle, E.J., Fasoli A., Gormezano, C., Gribov, Y. ,Gruber, O., Hender, T.C., Houlberg, W. , Ide, S. , Kamada, Y. ,Leonard, A., Lipschultz, B., Loarte, A., Miyamoto, K., Mukhovatov, V.,Osborne,T.H. , Polevoi1, A. & Sips, A.C.C. (2007) *Nuclear Fusion*, No. 47 (2007) S1-S17, ISSN 0029-5515

Snipes, J.A., Bora, D. & Kochanski, T.P. (1984). Initial Bolometric Measurements on Text. *Fusion Research Center Technical Report* , DOE/ET/53043-T2; FRCR-269, (Dec 1, 1984)

Stabler, A., McCormick, K. , Mertens, V. , Muller, E.R. , Neuhauser, J. , Niedermeyer, H. , Steuer, K.-H. , Zohm, H. , Dollinger, F. , Eberhagen, A. , Fussmann, G. , Gehre, O. , Gernhardt, J. , Hartinger, T. , Hofmann, J.V. , Kakoulidis, E. , Kaufmann, M. , Kyriakakis, G. , Lang, R.S. , Murmann, H.D. , Poschenrieder, W. , Ryter, F. , Sandmann, W. , Schneider, U. , Siller, G. , Soldner, F.X. , Tsois, N. , Vollmer, O. & Wagner, F. (1992) Density limit investigations on ASDEX Nuclear Fusion, *Vol.32, No.9, pp* 1557-1583, ISSN 0029-5515

Tahiliani, K., Jha. R., Gopalkrishana, M. V., Doshi, K., Rathod, V., Hansalia, C. & ADITYA team. (2009) Radiation power measurement on the ADITYA tokamak. *Plasma Physics and Controlled Fusion* Vol.51 (2009) 085004 pp 13, ISSN 0741-3335

TFR group, (1980). Bolometric Technics on TFR 600. *Journal of Nuclear Materials*, Vol. 93-94,Part-1, (1980) pp. 377-382, ISSN 0022-3115

Tokar, M.Z. (1995). Non-Linear Phenomena in Textor Plasmas Caused by Impurity Radiation. *Physica Scripta*, Vol.51, pp. 665-672, (1995), ISSN: 0031-8949

Wen, Y. & Bravenec, R.V. (1995). High-Sensitivity, High-Resolution Measurements of Radiated Power on TEXT-U. *Review of Scientific Instruments*, Vol.66, pp.549-552 (1995), ISSN *0034-6748*

Wesson, J. (2004), *Tokamaks* (Third Edition), Oxford University Press,0 19 8509227, OxfordYoung, K.M., Costley, A.E., Bartiromo, R., deKock , L., Marmar, E.S., Mukhovatov, V.S.,Muraoka, K., Nagashima, A., Petrov, M.P., Stott, P.E., Strelkov, V., Yamamoto, S., Bartlett, D., Ebisawa, K., Edmonds, P., Johnson, L.C., Kasai, S., Nishitani, T., Salzmann, H., Sugie, T., Vayakis, G., Walker, C., Zaveriaev, V., Perkins, F. W. , Post, D. E. , Uckan, N. A. , Azumi, M.,Campbell, D. J. , Ivanov, N. , Sauthoff, N. R. , Wakatani M., Nevins, W. M. , Shimada, M. & Van Dam J.(1999) Chapter 7: Measurement of Plasma Parameters. *Nuclear Fusion*, Vol. 39, No. 12 (1999), pp. 2541-2575, ISSN 0029-5515

Smart Bolometer: Toward Monolithic Bolometer with Smart Functions

Matthieu Denoual[1], Olivier de Sagazan[2], Patrick Attia[3] and Gilles Allègre[1]
[1]University of Caen Basse-Normandie, GREYC-ENSICAEN
[2]University of Rennes, IETR
[3]NXP-semicoductors Caen
France

1. Introduction

The content of this chapter refers to uncooled resistive bolometers and the challenge that consists in their integration into monolithic devices exhibiting smart functions. Uncooled resistive bolometers are the essential constitutive element of the majority of existing uncooled infrared imaging systems; they are referred to as microbolometer pixels in that type of application where matrixes of such elementary devices are used. Uncooled bolometers represent more than 95% of the market of infrared imaging systems in 2010 (Yole, 2010) and infrared imaging systems are required for more and more applications.

Mature industrial applications of uncooled IR imaging are non-destructive test and process control in production lines. Booming applications of uncooled IR imaging are in two main fields: security and environment. Application to security involves the conveyance security controls on one side and on the other side the emerging market of automotive security systems. In that case, infrared imaging is applied to the detection of pedestrians, animals or black ice on roads. At the present time, the high cost of IR imaging equipments prevents their broad distribution and restricts their market to luxury cars. The huge and dynamic market of automotive industry promises cost reduction in the next years. Besides, strengthening of safety norms such as Euro-NCAP in Europe will someday turn pedestrian detection systems into standard equipments. Environment is the second booming application field. It is driven by the environmental and ecological concern to track heat leaks in buildings and to allow for thermal budgeting of buildings. Driven by thermal regulation laws that limit the maximum power consumption of buildings, such as the RT 2012 in France, standard applicable from 2013, this field is to grow substantially in the coming years. According to recent market research, the volume of sale of uncooled infrared imaging system is to triple by 2015 (Yole, 2010) that is to say a 23% annual growth rate.

Such markets drive the research and development of uncooled infrared systems. Two main ways of development are investigated: (1) improvement of the bolometer pixel through appropriate choice of material and structure design, (2) optimization of the readout electronics. Amorphous silicon and other silicon based materials begin to challenge the historically dominant vanadium oxide (VOx) because their manufacturing is easier and cheaper. On the other hand, new packaging and microfabrication solutions such as through silicon via are considered for further integration and cost reduction. Concerning the second point, now, most of the readout electronics has been moved onto the chip where it is referred to as the readout integrated circuit (ROIC). The ROIC incorporates parallel column circuitry, consisting of amplifiers, integrators and sample-and-hold circuits with a column multiplexer which provides a single channel output. Most of the approaches today employ CMOS silicon circuitry for which the power dissipation is much lower than that of bipolar. Research works are still going on to improve this electronics, mainly for noise considerations (Chen et al., 2006; Lee, 2010; Lv et al., 2010). However, that electronics only enables the readout of the measured signal and cannot be directly derived to implement smart functions or operate the bolometer with feedback.

Independently from material or readout electronics, we address here another type of feature for uncooled bolometers that is the implementation of smart functions so as to derive "basic" uncooled resistive bolometers into smart bolometers. The smart qualification is not only a matter of readout circuit integration since it implies additional features compared to a bolometer pixel associated to conventional ROIC.

This chapter deals with smart bolometers according to the IEEE 1451.2 definition of smart sensors which states that smart sensors are sensors "that provide functions beyond those necessary for generating a correct representation of a sensed or controlled quantity". Test, identification and configurability are some examples of functions beyond conventional use, also called smart functions. Such smart functions contribute to an easier use of sensors and allow the sensors to take into account parameters discrepancies or evolutions. For instance, identification can be used to compensate for discrepancies between bolometers due to the process variations during the technological fabrication. Identification can, as well, allow the sensors to adjust to aging effects during their operating life. Combined with the configurability, the identification function makes possible to satisfy a large number of applications. The configurability takes advantage of the operation in a closed-loop mode to overcome the traditional trade-off between time constant and responsivity (Rice, 2000) and allows some flexibility in the choice of these characteristics.

The smart bolometers considered here consist in uncooled resistive bolometers associated to an electrical substitution means that enables the implementation of smart functions. The electrical substitution configuration chosen is the capacitively coupled electrical substitution (CCES) (Denoual et al., 2009a, 2009b). The demonstration of the closed-loop operation of resistive bolometers with this configuration has been performed with digital electronics implementation (Denoual et al, 2010). The introduction of digital electronics and control for feedback leads to new performance because of system linearization; in addition, it simplifies the implementation of smart functions. This configuration has the potential to allow for a fully integrated smart sensor, *i.e.* a monolithic smart bolometer. Such integrated smart bolometer is the ultimate goal of the work presented in this chapter and illustrated in Fig. 1 and Fig. 2. But before taking this rather large next step, it is important to understand the configurations that are available, and their ability to provide a higher level of intelligence and value to resistive bolometers.

For this purpose, prior to integration, high-level or top simulations are investigated as well as experimental prototyping with microcontroller units. One role of the top-simulation and modeling is to provide the designer with potential performance regarding various control algorithm strategies based on current or coming resistive bolometer technologies. Top simulations are desirable to validate the algorithms that implement the smart functions. Preferably, the validation of the algorithms should be performed in their operating context. These considerations guide the designer toward standard VHDL modeling technique as a solution. Such modeling technique exhibits indeed true interests including very fast simulation without convergence issues and the validation of algorithms in their operating context (Schubert, 1999; Staszewski et al, 2005). It is noticeable that this modeling technique is applied here to a multi-domain system involving optical, thermal and electrical signals.

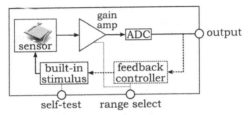

Fig. 1. General functional block diagram of a smart bolometer. Smart functions require a built-in stimulus generation. The feedback control and feedback path are optional and allow closed-loop operation of the bolometer.

Fig. 2. Integrated smart bolometer pixels. Line of monolithic smart bolometer pixels. Part of the conditioning and feedback electronics is integrated below the bolometer pixel.

Macroscale experiments with discrete components complete the knowledge acquired through the top-simulations. Those experiments involve resistive bolometer prototypes with capacitively coupled electrical substitution feedback means associated to microcontrollers that implements the smart functions for proof-of-concept demonstration. That is the current phase of development and one of the necessary steps to the next level: the monolithic smart bolometer.

This chapter is organized as follows:

After this introduction, the second section describes the smart functions to be implemented and stresses the need for built-in stimuli solutions. The third section includes top simulation and experimental results. The contexts of those results are presented; especially the modeling technique for the top simulation and the experimental set-up for the macroscale and discrete component based results are described. Based on these results, the expected performance of a fully integrated smart bolometer is explored. The ultimate capabilities of smart bolometers will be limited only by the performance of the integrated electronics and the imagination of the designer. The fourth chapter highlights a particular solution for the integration of the bolometer and its associated electronics for the implementation of smart functions that would enable the realization of lines or matrixes of pixels for smart infrared imaging systems.

2. Bolometer with smart functions

Referring to literature, functions performed by smart sensors in that role include correcting for environmental conditions, performing diagnostic functions, and making decision (Frank, 2000). The smart functions implemented in the smart bolometer are described in this section as well as the means required for their implementation.

Before getting further into the description of the smart functions and their implementations, some notations and vocabulary are defined.

An uncooled resistive bolometer converts absorbed infrared (IR) radiation into heat, which in turn changes the resistance of a sensing resistor. The sensing resistor or thermistor is current biased. A bolometer can be modeled as an IR-sensitive element of thermal mass C_{th} linked via a thermal conductance G_{th} to a substrate acting as a heat sink (see Fig. 3).

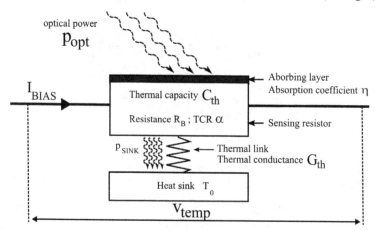

Fig. 3. Schematic of a resistive bolometer.

The performance of the bolometer is characterized by figures of merit such as its responsivity (R), the temperature coefficient of resistance (TCR or α, positive or negative depending on the material) of the temperature sensing resistor, its specific detectivity (D^*) and its effective time constant ($\tau_{eff}=C_{th}/G_{eff}$, G_{eff} is the effective thermal conductance and depends on G_{th}, that takes into account the electro thermal effect (Richards, 1994)).

The responsivity (R) describes the variations of the output voltage signal (v_{temp}) depending on the IR input radiation (p_{opt}) and it is expressed by the transfer function of the bolometer as follows:

$$R(\omega)[V/W] = \left| \frac{v_{temp}(\omega)}{p_{opt}(\omega)} \right| = \frac{\alpha \eta I_{BIAS} R_B}{\sqrt{\left(G_{eff}^2 + \omega^2 C_{th}^2\right)}} \tag{1}$$

where η is the absorption coefficient of the absorption layer of the device, I_{BIAS} is the bias current, R_B is the bolometer resistance (sensing resistor).

The trade-off between time constant (τ_{eff}) and responsivity (R) appears in the dependence of those parameters according to the thermal conductance. A low G_{eff} is required to improve the responsivity but this negatively increases the time constant.

2.1 Investigated smart functions

The smart functions of the smart bolometer described here are of two kinds: diagnostic functions and correction functions. A third type of smart functions is decision making which is not addressed here. Among the diagnostic functions, the first one is self-test. The self-test feature allows the verification of the thermal and electrical integrity of the bolometer at any time during its operating life. It provides the user with a qualitative result that informs whether the bolometer is working or not. The second diagnostic function is self-identification. The self-identification feature is more complex than self-test. The self-identification allows the characterization of the sensor and its associated electrical circuitry. This feature can be used at any time for monitoring the aging of the device and for deciding if a calibration is required. This feature is useful if closed-loop mode operation of the bolometer is considered in order to extract the forward path parameters (bolometer and its conditioning electronics) for the evaluation of the parameters of the controller that would drive the feedback path (Fig. 1.). Fig. 4. depicts the functional block diagram of an adaptive identification algorithm. Adaptive algorithms are interesting in that they run in real-time and do not require huge memory means since a few parameters and a few coefficients are stored. The identification principle is to iteratively adjust the parameters of the model to make the predicted output of the model converge towards the output of the forward path. The adjustment is performed according to the stimulation input signal and the error between the predicted output of the model and the current output. This convergence enables the extraction of estimated parameters representing the device, especially the time constant, the DC responsivity and the thermal characteristics of the bolometer.

Self-identification refers to identification using a built-in stimulus. The same type of identification algorithms can be used for calibration, in that case external optical stimuli are used and identification results are used to derive coefficients stored in a calibration table.

After diagnostic functions, the developed smart bolometer implements a correction function that is range selection. Open-loop and closed-loop operation modes should be distinguished. In open-loop operation mode, the input range can be modified by changes of the gain of the conditioning electronics or more rarely through the current bias of the sensing resistor of the bolometer. In closed-loop mode, the input range is selected by the gain of the controller and the gain of the feedback. The closed-loop mode, in addition, allows input range selection around a user defined operating point.

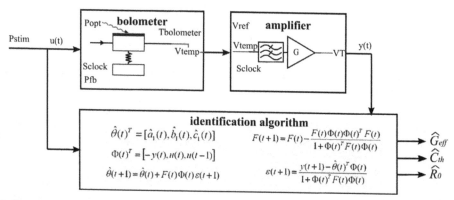

(^ indicates estimation)

Fig. 4. Identification functional block diagram. With $u(t)$ the stimulation input (Pstim), $y(t)$ the open-loop system output, $\hat{\theta}(t)$ the parameters of the model, $\Phi(t)$ the observation vector, $\hat{\theta}(t)^T \Phi(t)$ the predicted output, $\varepsilon(t)$ the error between the system output and the predicted output, $\hat{G}_{eff}, \hat{C}_{th}, \hat{R}_0$ the extracted estimated parameters representing the device.

Considering lines or matrices of pixels, the identification smart function associated to closed-loop operation is a way of compensation of the spatial noise caused by the bolometer resistance dispersion due to fabrication process. Bolometer pixels individually operating in closed-loop mode would be able to compensate this spatial noise after external calibration or built-in calibration thanks to built-in input stimulus.

2.2 Requirements

Addressed implicitly in the previous sub-section, a built-in stimulus means is the first requirement to implement self-test or self-identification functions, the second one being a digital core for the implementation of the algorithms of the smart functions. A built-in stimulus implies being able to stimulate the sensor with a self-generated signal in the same manner as an external optical power stimulus would do. The first and straightforward solution is to use an optical power source. The major drawback of this solution is the integration limitation, especially if lines or matrixes of pixels are considered. The second solution relies on the electrical substitution principle (Rice, 2000) also referred to as electrical equivalence principle (Freire et al, 2009). This principle states that Joule heating electrically produced can be used to equivalently stimulate the sensing resistor of a bolometer compared to optical incoming power.

The availability of a built-in stimulation source can lead fairly directly to the closed-loop operation of the sensor. Indeed, only a controller has to be inserted in a feedback loop. The advantages obtained through closed-loop operation of bolometers detailed in (Denoual & Allègre, 2010) are rapidly recalled here. For instance, closed-loop operation mode enables to increase the bandwidth and the range of the measurement. Closed-loop mode makes it possible to work around a user-defined operating point. This feature is important for the input range selection around a user-defined operating point.

In practice, three ways exist for the realization of an electrical substitution stimulus and its extension for closed-loop operation of the bolometer. There are all based on electrical Joule heating generation and differ depending on whether the Joule heating is generated onto the sensing resistor of the bolometer or at its vicinity. The Joule heating is either produced by a current or a voltage. From now, voltage stimuli for the Joule heating generation are considered. The three implementations of electrical substitution are schematically represented in Fig. 5(b-d).

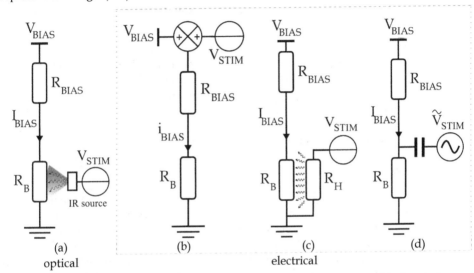

Fig. 5. Built-in stimulus means. (a) optical power source implementation. (b) electrical substitution (ES) configuration type 1. (c) ES type 2. (d) ES type 3. With R_{BIAS} the bolometer biasing resistor, R_B the sensing resistor of the bolometer, R_H an additional resistor for Joule heating, V_{BIAS} the biasing voltage, V_{STIM} the voltage stimulus, I_{BIAS} the DC bolometer bias current, and i_{BIAS} the variable bias current.

In type 1 implementation, Fig. 5(b), the voltage stimulus, V_{STIM}, is added to the bias voltage V_{BIAS} (Freire et al, 2009). The voltage stimulus produces Joule heating into the sensing resistor of the bolometer. The voltage across the sensing resistor of the bolometer changes according to the resistance variations of the sensing resistor due to both Joule heating variations and bias current, i_{BIAS}, variations. In that case, the thermal and electrical operating points of the bolometer are intimately linked. This particular link makes the operation in closed-loop mode tricky and commonly yields to stability issues and does not allow taking advantage of the bandwidth increase due to closed-loop operation as pointed out in (Williams, 1990).

The type 2 implementation, Fig. 5(c), uses an additional resistor (R_H) as a heater to produce the Joule heating stimuli (Rice et al, 1998). It enables the separation between the electrical and the thermal operation points at the expense of an additional resistor. This additional resistor has to be close to the sensing resistor, which adds a constraint to the design of the bolometer. Operation in closed-loop mode using this configuration is successfully demonstrated in (Rice et al, 1998; Allègre et al, 2007).

The type 3 implementation, Fig. 5(d), exhibits the advantages of the type 1 and of type 2 implementations without their disadvantages, *i.e.* no additional resistor but separation of thermal and electrical operating points. This implementation is compatible with existing resistive bolometer without any material or design modification but still allows separating the electrical and thermal operating points (Denoual et al, 2009). The operating points are separated according to frequency basis. The bial voltage signal, electrical operating point, is a low frequency signal while the voltage stimulus signal for the thermal operating point is a high frequency signal, typically tens of MHz. No additional heat source is needed, and stability issues are fixed.

All this built-in stimulus implementations can be derived to operate the bolometer in closed-loop mode (Denoual & Allègre, 2010). The four built-in stimulus solutions illustrated in Fig. 5 may be used to implement smart functions and derive a smart bolometer. However, the ease of use and the flexibility of the type 3 implementation, Fig. 5(d), make it the best suitable candidate for smart bolometer integration despites its extra electrical circuitry compared to the other two electrically based solutions. This is demonstrated in the next section through simulation results and experiments on macroscale bolometer.

3. Actual work

Before getting into the long and expensive design and fabrication process of the monolithic smart bolometer, it is essential to perform validations to determine the feasibility and to guide the design. Our validation procedure goes through top simulations and macro-scale experiments with discrete components. Some results of both simulations and experiments are presented in this section to demonstrate the potentialities of integrated smart bolometers. As mentioned in the previous section, the integrated smart bolometer is based on the capacitively coupled electrical substitution (CCES) technique for built-in stimulus generation and closed-loop operation. Before getting into the results of simulations or experiments, the key features of the CCES principle are recalled; more information can be found in (Denoual et al, 2009; Denoual et al, 2010; Denoual & Allègre, 2010).

Capacitively Coupled Electrical Substitution (CCES)

The principle of the closed-loop mode operation using the electrical substitution (ES), also involved in the CCES technique, is depicted through Fig. 6 and Fig. 7, and described hereafter. With no external optical power applied, the thermal operating point (PBIAS) is set by the VPbias value. Typically the bolometer is heated up a few Kelvin, at temperature T_{BIAS}, over the room temperature, T_0. When an external optical power, P_{opt}, is applied, a change in the resistance of the bolometer sensing resistor is sensed, and consequently the output signal of the controller changes, decreasing the average power of the feedback Joule heating, P_{fb}. The net effect is to produce a Joule heating to counterbalance the externally applied optical power on the bolometer so as to keep the temperature of the sensing resistor of the bolometer constant at T_{BIAS}.

Among ES techniques, the CCES technique presents several distinct characteristics.

The capacitively coupled electrical substitution dissociates the electrical and thermal operating points according to a frequency basis as already mentioned. High frequency modulated signals are used for the voltage applied to the sensing resistor for the heat feedback. Since the feedback is high frequency, the feedback signal does not end up in noise signals in the bandwidth of the sensor.

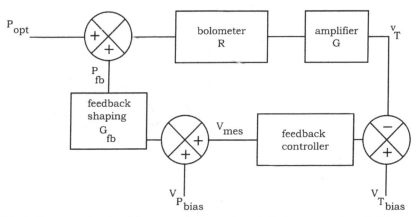

Fig. 6. Block diagram of a closed-loop implementation of a bolometer using electrical substitution (ES). G is the amplifier gain and G_{fb} is the gain of the feedback shaping module. V_{mes} is the output in closed-loop mode.

This implementation can be applied to any kind of uncooled resistive bolometer. The digital implementation, involving pulse width modulation (PWM) or Sigma-Delta modulation, in general terms pulse coded modulation (PCM), enables the linearization of the feedback path as well as a direct digital output power reading (Denoual et al, 2010). Clocking for this type of system is typically about 1-10 MHz. Fig. 8 depicts one example of such digital implementation.

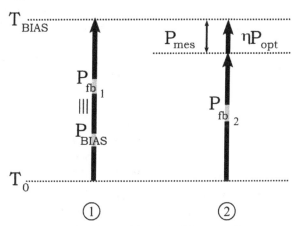

Fig. 7. Electrical substitution principle used for closed-loop mode operation. ① without external optical power, the thermal operating point is set by the bias power corresponding to the V_{Pbias} voltage. ② with external optical power, the total amount of power dissipated onto the bolometer is kept constant thanks to the variations of the feedback power.

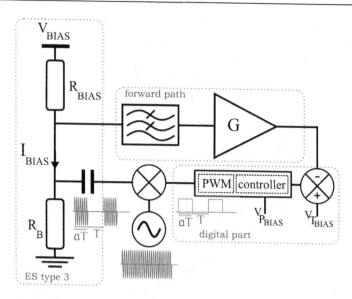

Fig. 8. Schematic of a digital implementation of the capacitively coupled electrical substitution technique for closed-loop operation of a resistive bolometer involving a PWM modulation with a duty cycle a.

3.1 Top-simulation and modeling

Generally speaking, the main specification for the environment used for the top-simulation of a smart sensor is its ability to validate the algorithms implementing the smart functions in their operating context, *i.e.* including the sensor and its associated electronics. Few simulation environments address this specification because it implies mixed simulations (analog/digital) and multiphysics simulations, here optical, thermal and electrical. Environments commonly used address only one of each simulation type; for instance ANSYS® or COMSOL® for multiphysics simulation widely used for MEMS sensor design and simulation, SPICE-like environments for analog simulation and NCSim® or ModelSim® environments for digital simulation.

Spice-like environments may be used for top-simulation of system combining electronics and sensors through modeling of the sensor with equivalent electrical circuits (Jones et al, 2003). However, Spice-like simulators are not suitable for the validation of algorithms because the algorithms implemented on the digital part cannot be run. Only snapshots corresponding to specific and static configurations can be tested.

Long time simulations or huge processing requirements prevent the usage of coupled or mixed environments for the type of simulation considered here.

Modeling using VHDL-AMS or Verilog-AMS has been proposed for the modeling and for multi-domain simulation of MEMS devices in their functional environment (Chapuis et al, 2008). But as far as top validation is concerned, *i.e.* test of embedded algorithms in their operating context, these techniques exhibit huge simulation times that are not compatible with the validation of the algorithms and usually suffer from convergence issues. Besides, the set-up of that kind of simulations is often tricky.

Matlab with appropriate modeling of the analog blocks of the system can be used for this kind of top-simulation, but no direct link exists between the electrical schematic and the Matlab schematic. This is a disadvantage when the design of the integrated circuit and monolithic device is considered.

The simulation environment chosen here is purely digital and enables fast simulation without convergence issues. It is associated with a modeling technique already applied for mixed integrated electronic circuits such as phase-locked loops (Schubert, 1999; Staszewski et al, 2005) and uses the standard VHDL as a description language (Denoual & Attia, 2011).

Advantages of the used modeling technique come from: (i) the event-driven nature of the simulation using purely digital environment and (ii) the properties of the standard VHDL language. The event-driven nature of the simulation results in drastically shorter simulation times compared to time-driven simulation using for instance Spice (Zhuang et al., 2006). This modeling technique does not suffer from the convergence issues usually observed with other techniques because it uses digital simulation environments (ModelSim, NCSim,...). Finally, the standard VHDL syntax with user's defined types enables implicit connectivity check between the parts of the designed system since in VHDL two connected signals must have the same type.

3.1.1 Modeling technique

The basic principle of this modeling technique is the discretization of the analog parts of the design. An appropriate modeling of the analog parts of the design is desirable to overcome the problematic induced by different time scales. This type of problematic exists in this work involving low frequency thermal phenomena (< kHz), and high frequency electrical signals. All the elements of the system are modeled using standard VHDL language. Those elements are schematically presented in Fig. 9 corresponding to the testbench. The testbench also includes the optical power stimuli and the stimuli process. In this example, the smart function simulated is open-loop identification.

In the case of this study, the modeling of the analog blocks, illustrated in Fig. 10, is performed in 3 steps: 1- definition of analog transfer function of the block; 2- discretization of the analog model; 3- VHDL description of the discretized model.

The model for the bolometer corresponds to the VHDL transcription of the discretized transfer function (1). The discretization is achieved using the bilinear transformation. The conversion process is divided into two consecutive processes: the thermal process dealing with power inputs and temperature, and the electrical process corresponding to the temperature measurement. This divided structure would enable to take into account the electrothermal feedback phenomenon (Richards, 1994) of the bolometer itself if necessary in the simulation. At the present time, this phenomenon is taken into account through the use of the effective thermal conductance (G_{eff}) rather than the physical thermal conduction (G_{th}).

The models for the amplifier block merely consists in a gain since the bolometer voltage output signal is in the bandpass of the filter.

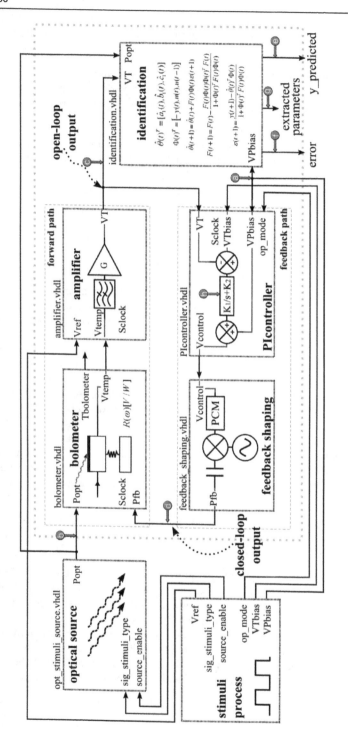

Fig. 9. Complete Testbench. The circle marked arrows indicate the signals observed in the result of simulation presented in this section. (a) V_{Pbias} is the voltage for the built-in electrical stimulus; (b) P_{opt} represents the optical input signal; (c) V_T is the output in open-loop operation mode; (d) P_{fb} is the feedback power considered in the simulations as the output in closed-loop operation mode; (e) is the predicted output evaluated by the identification algorithm; (f) is the error between the predicted output and the real output; (g) corresponds to the extracted parameters; (h) represents the controller parameters.

The model for the controller block in this case implements the equations of a digital proportional integral (PI) controller. The operation mode input, op_mode, enables to choose between open or closed-loop operation modes.

The model for the feedback shaping block consists in a gain and saturation limitations corresponding to the PWM modulation. It linearly gives the feedback power (P_{fb}) according to the duty cycle a of the PWM modulated control signal (Vcontrol) following $P_{fb} = a \cdot G_{fb}$, with G_{fb} the gain of the feedback shaping block. G_{fb} is a function of the amplitude of the carrier ($V_{carrier}$) and the resistance of the sensing resistor of the bolometer (R_B). The high frequency carrier that translates the feedback bandwidth is not taken into account. Pragmatically, no signal at this frequency is generated for the simulations, only the feedback duty cycle is considered.

The optical source block generates stimuli with parameterized frequency, amplitude and shape.

The identification block implements a least-mean-square adaptive fitting algorithm which role is to extract parameters in order to optimize the feedback controller and/or to monitor the aging of the device; it corresponds to the functional block diagram of Fig. 4. Detailed explanation of this algorithm and of the parameters involved can be found in (Landau, 1988; Ljung, 1999).

The complete modeling of the bolometer and its associated conditioning and feedback electronics as well as the modeling of the external optical source enable the validation of algorithms implemented in digital parts. Identification is an example of such algorithms that can be validated using this modeling technique and digital simulation tools.

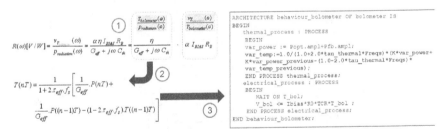

Fig. 10. Standard VHDL modeling procedure. ① analog model of the transfer function of the resistive bolometer, ② discretization of the analog model, ③ VHDL transcription of the difference equations.

3.1.2 Top-simulation results

This section illustrates the type of simulation that can be performed with this modeling technique. The simulations were performed using the ModelSim® Altera 6.3 Quartus II 8.1 software, i.e. a purely digital environment. The figure Fig. 11 shows the three configurations of the simulation sequence performed and illustrated in Fig. 12. First an open-loop operation sequence, Fig. 11(1), illustrates the time constant and magnitude of a typical response of the bolometer when exposed to an incoming square shape optical power P_{opt}. The second sequence, Fig. 11(2), corresponds to a built-in random stimulation to perform the identification procedure. The adaptive least-mean-square algorithm illustrated in Fig. 4 is implemented using standard VHDL to extract the characteristic parameters of the bolometer while the stimulus is applied. The duration of the identification procedure depends on the time constant of the bolometer. To achieve the convergence of the identification parameters, an identification duration of several hundred times the bolometer time constant is required. A 10 ms time constant bolometer leads to an identification procedure of a few seconds. Finally, in the last sequence, Fig. 11(3), the bolometer operates in closed-loop mode thanks to a controller adjusted with the extracted parameters at the end of the identification procedure. A standard factor 10 for the time constant reduction is applied in this simulation. One should notice that the time scale is not mentioned since all the timing parameters (sampling frequency, PWM frequency, carrier frequency) are adapted to the bolometer and therefore scale up or down with the time constant of the bolometer. The useful time reference is the time constant of the bolometer in open-loop.

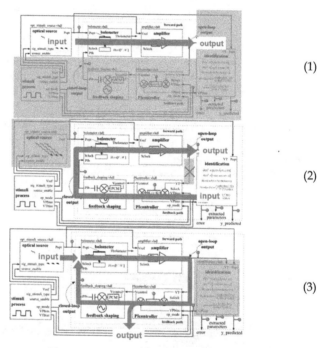

(1)

(2)

(3)

Fig. 11. Simulation sequences. (1) open-loop mode operation. (2) identification procedure mode. (3) closed-loop mode operation. According to the operation mode, the location of the output differs. Note that input may be either external and optical (1,3) or internal and electrical (2).

Simulations in either open-loop or closed-loop are performed without convergence issues within a few seconds. This enables fast parameters optimization for the control through series of simulations. It shall be mentioned that equivalent simulations with analog environments (spice-like, coupled or mixed simulators) would take hours if no convergence problems occur. A simulation result is represented in Fig. 12. This simulation underlines the ability of the modeling technique to validate algorithm supporting smart functions in their operating context by top simulation.

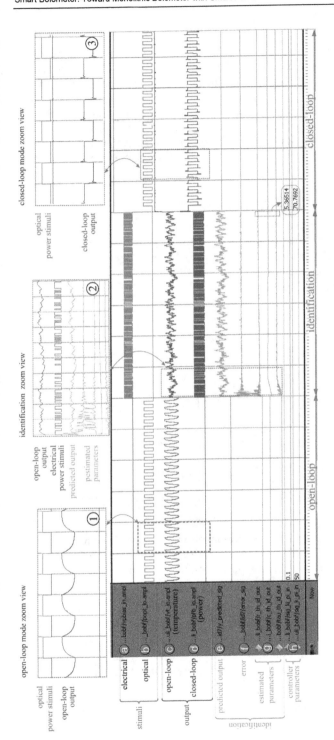

Fig. 12. Example of top simulation chronogram. Three consecutive operating sequences are represented: 1- open-loop mode operation, 2- identification process in open-loop mode, 3- closed-loop mode operation. Insets illustrate the typical signals in each case. During the open-loop and closed-loop operating phases, the stimulus is an optical square shape signal. During the identification process, the stimulus is electrical. A pseudo-random binary sequence is used in order to optimize the identification process. The identification process allows the estimation of the thermal parameters of the bolometer. It consequently enables the evaluation of the parameters of the controller in order to reach desired performance, here a time constant in closed-loop 10 times smaller than the one in open-loop and a damping of 0.7 for stability and speed of response reasons. During the closed-loop operation, the amplitude of the substituted electrical power is determined by negative feedback within the controller of the feedback path. When the feedback power compensates the incoming optical power, the temperature of the sensing element of the resistive bolometer is constant. Consequently, the feedback power is a direct reading of the incoming optical power.

3.2 Macro-scale and discrete components experiments

Macro-scale experiments are important in that they validate the models of the top-simulations presented in the previous section. The top-simulations validate the system functionality and the smart function algorithms during the design phase, whereas the experiments exhibit additional non-simulated factors such as noise or nonlinearities and therefore exhibit the performance of the system. This section presents a macro-scale and discrete components set-up and some results of characterization performed with it. The major results are the successful reduction of the time constant of the system by at least two orders of magnitude, and the practical demonstration of the self-test and range selection smart functions.

3.2.1 Description of the macro-scale set-up

The set-up, depicted in Fig. 13, consists in a pair of resistive bolometers, R_B, and a digital implementation of the CCES technique for the built-in stimuli generation and the closed-loop mode operation. The resistive bolometers are connected in a bridge arrangement with one sensing resistor in each leg of the bridge (R_{BIAS} and R_B). Only one is exposed to the optical power stimuli for differential measurement.

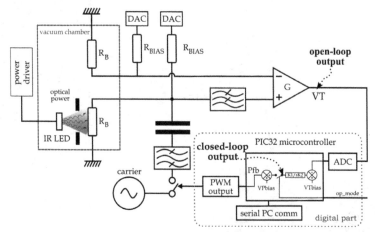

Fig. 13. Set-up for the macro-scale and discrete components experiments. Two resistive bolometers, R_B, under vacuum, are used for differential measurement. Only one is exposed to the infrared optical power from a LED. The exposed bolometer in operated in closed-loop mode through a digital implementation of the CCES technique.

The bolometers are macro-scale suspended bolometers made of 60 μm thick, 0.8 cm large, 6 cm long suspended PolyEthylene Naphtalate (PEN) membranes covered with aluminum metallic layer. The aluminum layers are patterned to form the 700 Ω resistors used as sensing resistors. The active optical power absorption area is 0.5x0.5 cm². Global characteristics of those bolometers extracted from measurements at 20 mTorr and room temperature are:

TCR = 2.3×10^{-3}/K; G_{eff} = 550 μW/K, and τ_{eff} = 110 s (cut-off frequency F_c=1.45 mHz).

The huge time constant, τ_{eff}=110 s, is a consequence of the macro-scale nature of the bolometer.

A microcontroller (PIC32) implements the closed-loop controller and the embedded algorithms. Two digital-to-analog converters (DAC) drive independently the voltage of the legs of the bridge to set the electrical bias point (Denoual et al, 2010). The operation mode input (op_mode) is a logic signal that selects between open-loop or closed-loop operation mode. The PWM modulated output signal for the feedback results in a linear dependency of the heat power feedback on the control voltage (P_{fb}). The frequency of the carrier is fixed at 1 MHz while its amplitude changes between 100 mV and 200 mV depending on the measurement range choice.

All the experiments are led at room temperature and under a primary vacuum of 20 mTorr to limit convection effects.

The external optical power stimuli source is an infrared LED with a 1 mW power at λ=1.45 μm. The amplitude of the optical power is modified by duty cycle modulation (with duty cycle d) of the voltage of the power driver as in (Denoual et al, 2010) and illustrated in Fig. 14. The frequency, $1/T_{PWM}$, of the pulse modulated signal is 10 kHz, *i.e.* several orders of magnitude over either the open/closed-loop bandwidth. Consequently, the bolometer in open-loop or closed-loop operation modes merely responds to the average value of the applied power as illustrated in Fig. 14.

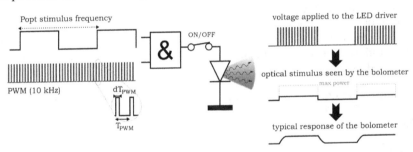

Fig. 14. IR LED supply principle. A double modulation enables to generate square shape optical stimulus with controllable amplitude.

3.2.2 Macro-scale and discrete components experimental results

3.2.2.1 Time constant reduction

The first set of experiments, illustrated in Fig. 15, aims to demonstrate the possibility of reducing the time constant by at least 2 orders of magnitude.

During this experiment, the bolometer successively operates in open or closed-loop mode. During optical power ON sequence, 0.1 Hz frequency optical stimuli are applied from the infrared LED of the set-up. During power OFF sequence, no optical power is applied.

The open-loop output signal during the open-loop operation phase illustrates the time constant of the system in open-loop through its typical first order response. The system in open-loop acts as a low-pass filter with a cut-off frequency of 1.45 mHz (τ_{eff}=110 s) that filters the 0.1 Hz frequency input optical stimuli and only responds to the activation of the

ON sequence of the optical power. On the contrary, during closed-loop operation zoomed in the upper view of Fig. 15, as the bandwidth of the system is increased, the closed-loop output signal is able to follow the 0.1 Hz variations of the optical input stimuli.

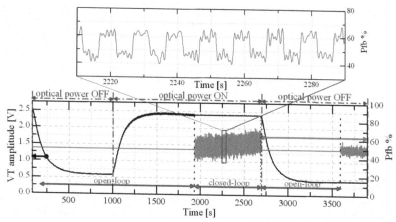

Fig. 15. Experiment illustrating the time constant reduction in closed-loop operation mode. The main graph represents the recorded output signals in open-loop (VT) and closed-loop (Pfb). The open-loop output VT is in Volt (left axis) while the closed-loop output is in percentage of the feedback power full-range (right axis). The optical stimulus sequence is ON [0·s], OFF [0 s ; 1000 s], ON [1000 s ; 2700 s], OFF [2700 s ; end]. During the ON sequence of the optical stimulus, in the center of the graph, the operation mode is successively open-loop and closed-loop.

By operating the system in closed-loop mode, the bandwidth is increased from 1.45 mHz to more than 3 Hz. This result highlights an improvement of the system bandwidth by more than 200.

3.2.2.2 Smart function validation: Self-test

Although this function is rather basic, it may be useful for the user as a basic diagnostic function to answer the crucial question: "is the bolometer and its associated electronics working or not ?". This feature requires the existence of a built-in stimulus input. The signal V_{Pbias} that sets the thermal working point of the system is used for this built-in stimulation (see Fig. 1, Fig. 6 and Fig. 13). Typically, self-test is activated by the user with a logic high level on the self-test input pin. During the logic high level, a Joule heating is applied onto the sensing resistor of the bolometer equivalent to approximately 20% of full-scale optical input power, and thus a proportional voltage change appears on the output signal, either open-loop or closed-loop output depending on the configuration test. When activated, the self-test feature exercises both the entire thermal structure and the electrical circuitry, and in addition in closed-loop mode the feedback path. The results presented in the Fig. 16 illustrate this functionality both in open or closed-loop. In open-loop mode, the pulsed stimuli on V_{Pbias} result in pulsed response at the output indicating that the bolometer is working properly. In closed-loop mode, the pulsed stimuli on V_{Pbias} result in pulse response at the closed-loop output while the feedback signal, P_{fb}, is maintained constant so as to keep the thermal working point constant. The different time scales in each cases illustrates once again the time constant reduction in closed-loop mode.

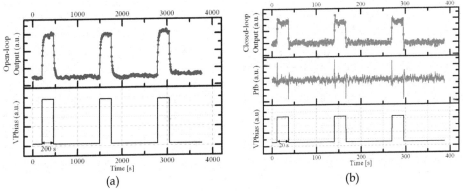

(a)   (b)

Fig. 16. Self-test response in open-loop (a) and closed-loop (b) operation modes. Pulses on V_{Pbias} generate Joule heating onto the resistance of the bolometer through the feedback shaping electronics so as to stimulate a response from the device.

3.2.2.3 Smart function validation: Range and operating point selection

The next experimental results illustrate the range selection ability as well as the possibility of operating around a user-defined operating point. Fig. 17(a) shows the measured signal in open-loop mode as a function of the input power while Fig. 17(b) shows the measured power in closed-loop mode as a function of the input power. These transfer functions are obtained with infrared LED source stimulation in the bandwidth of the device (10 mHz in the open-loop mode configuration and 100 mHz in closed-loop mode configuration).

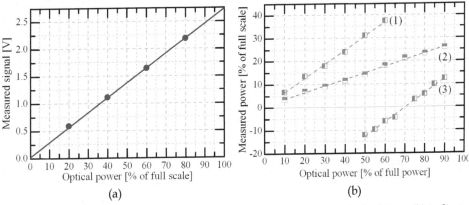

(a)   (b)

Fig. 17. Transfer functions in open-loop (a) and closed-loop (b). The dashed lines (b) indicate the two different slopes of the transfer functions in closed-loop mode. (1) and (3) exhibit a slope twice as big as the one of (2). (3) is the transfer function in closed-loop mode around 70% optical input power.

In open-loop the measured signal (V_T) is a function of the input optical power, responsivity of the bolometer and the gain of the forward path amplifier. Therefore, the transfer function can be adapted to various incoming power ranges by modification of the gain of the amplifier. The modification of the responsivity through the bias current is not relevant because the signal-to-noise ratio is negatively impacted if the responsivity is decreased. The responsivity has to be as high as possible according to the fabrication technology.

In closed-loop mode, the output signal is a function of the input power and the feedback shaping gain. The overall measurement range is then given by the pulse coded modulation range, the ADC range and the feedback gain. Therefore the measurement range of the device may be easily modified to measure lower power or higher power optical stimuli by respectively increasing or decreasing the feedback shaping scale factor. G_{fb} stands for the feedback shaping gain (see Fig. 6) involving the carrier voltage amplitude, the PCM voltage amplitude and the filtering amplification gain. The feedback shaping scale factor specifies the voltage change of the output per power unit of applied optical power. The selection of the operating point achievable in closed-loop mode also enables to move the operating point of the transfer function of the bolometer around a user-defined operating point, in order for example, to measure optical power variations in an input optical signal of given mean power value.

These possibilities in closed-loop mode are illustrated in the graph in Fig. 17(b). Two transfer slopes resulting in two different measurement ranges are shown. In addition the transfer curve (3) of Fig. 17(b) shows the transfer of the system to optical power variations around an optical power mean value of 70% of the maximum IR LED power.

Such control of the measurement range shall allow the implementation of algorithm that dynamically adjusts the scale to prevent saturation and optimize the resolution.

3.3 Experimental results section conclusion

The results section experimentally illustrates that a bolometer can be speed up by 2 orders of magnitude. Such time response improvement gives freedom in the design constraint of bolometers and especially for the conventionally responsivity/time constant tradeoff. For example, standard micro-bolometers designed for imaging are designed to meet the frame image refreshment rate of 20 Hz. This constraints the upper time constant limit for these devices. On the other hand, the thermal capacitance of these micro-bolometers cannot be boundlessly reduced because of material stiffness required to ensure self-sustention. Consequently, time constant consideration rather than responsivity optimization determines the thermal conductance. The demonstrated two orders of magnitude reduction of the time constant opens the door for bolometers two orders of magnitude faster than existing ones (200 µs compared to 20 ms) or bolometers exhibiting the same time constant but with a two order of magnitude higher responsivity. In both cases, such devices characteristics are not achieved yet. Currently the commercially available IR image sensors have detectivity, proportional to responsivity, of the order of few 10^8 $W^{-1}.cm.Hz^{1/2}$ and image refresh rate of 10-30 frame/s. Two orders of magnitude improvement means detectivity and refresh rate reaching 10^{10} $W^{-1}.cm.Hz^{1/2}$ and several 1000 frames/s respectively. New application fields of infrared-monitoring in chemical and entertainment fields for instance will be then possible, while the existing applications fields will be reinvigorated. Faster frame rate of IR imager used for process control will allow increasing the speed of production line; the economic impact is direct. In automotive safety applications, higher image data rate with optimized algorithms will result in faster and more reliable detection. For thermal budgeting of buildings, improved detectivity will enhance diagnostics. In all cases, the extra smart functionalities facilitate the use of IR imagers.

4. Monolithic 3D smart bolometer

Results shown in the previous section, demonstrate the possibly of realizing bolometer exhibiting smart functions such as self-test or range selection. The integration on one single chip of such device would yield to monolithic smart bolometer following the example of integrated accelerometers. Even though such devices might be interesting and competitive compared to thermopile-based IR detectors applied to distance temperature measurement (Texas, 2011), the real goal application of such devices is imaging. Imaging requires lines or matrixes of pixels and consequently adds geometrical constraints to the integration of smart bolometer pixels. The CCES technique enables closed-loop operation and smart functions implementation with built-in stimuli at the cost of an electrical circuitry bigger than the readout integrated circuitry of standard bolometer imaging devices. Consequently, the integration with existing topology and fabrication processes of the CCES circuitry under each pixel of an imaging device seems quite challenging. However, emerging micromachining technologies might come as a rescue to provide a solution. This solution is discussed in this prospective section.

The integration issue comes from the lack of space under the pixel to integrate the CCES with actual planar conventional technologies, *i.e.* 2D technologies. Adding one dimension gives some more space and design freedom. A 3D geometry, as illustrated in Fig. 18, dissociates the sensing area of the pixel from the area needed for the electrical circuitry implementation. While the electrical circuitry is designed onto the device surface as it is common, the sensing area of the pixel is realized vertically in the bulk of the substrate.

Such 3D design is made possible by the emerging 3D micro-fabrication technologies. Fig. 19 illustrates the typical geometries and shapes available with submicron deep etching process. Vertical 50 µm walls in depth with few hundred nanometer thickness are reported using this etching process (Mita et al, 2006; Hirose et al, 2007). Such characteristics are fully compatible with the realization of the vertical sensing area, the resonant optical cavity for absorption enhancement and the folded legs behind, all together constituting the bolometer pixel. The sensing area is typically 30 µm to 50 µm wide and few hundred nanometers thick. The thickness of the sensing area is usually reduced in order to reduce the thermal capacitance of the bolometer pixel. The operation in closed-loop mode can release this constraint because of the bandwidth increase (time constant reduction) achievable in closed-loop mode. The resonant optical cavity, which enhances the absorption of incoming optical power, consists of a $\lambda/4$ space between the sensing area and a reflective layer. Taking into account the mid-infrared range object of such devices $\lambda/4$ is a few microns distance fully compatible with the submicron deep etching process. Usually, efforts to increase the fill factor include reducing the size of the contacts and can use buried legs on an intermediate level between the level of the sensitive area and that of the substrate. However the buried legs must not interfere with the increased absorption provided by the resonant optical cavity

(Kruse, 2001). In the proposed 3D configuration, the folded legs do not interfere with the resonant optical cavity. The folded legs can be lengthened to reach a target thermal conductance without negatively impacting the fill factor. Considering the etching performances of the submicron deep etching process, the legs can be very narrowly folded. Hence, the distance between the fold of the legs would not be limited by the fabrication technology but by thermal transfer at submicron scale. Indeed, at nanoscale, the thermal transfer mechanisms change and radiative heat exchange mechanism increases as the dimension between surfaces decreases below 200 nm because of surface phonon polariton mediated energy transfer (Rousseau et al, 2009; Shen et al, 2009). The space between the fold of the legs should not be below 200 nm to ensure that the thermal conductance of the pixel is only mediated by the conduction mechanism. Taking this into account and the actual pixel geometry and thermal conductance (10^{-6}-10^{-7} W/K), the same thermal conductance could be achieved within a 30 μm space behind the pixel.

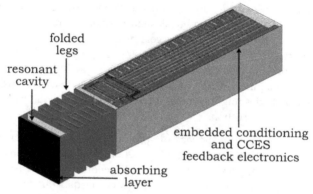

Fig. 18. 3D bolometer pixel. The pixel is micro-machined vertically into the substrate while the conditioning and feedback electronics is on-plane.

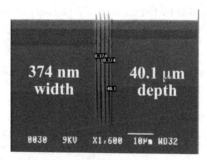

Fig. 19. SEM picture of submicron deep trenches in silicon. (*Courtesy of Mita Lab., University of Tokyo*).

The 3D bolometer pixel and its association into matrixes to form a smart IR retina, depicted on Fig. 20, as defined with Dr. Yoshio Mita of the university of Tokyo, represents a technological breakthrough in the design, the fabrication and the use of uncooled resistive bolometers for infrared imaging; a technological breakthrough leading to faster, more sensitive and smart infrared imaging devices. The 3D bolometer pixel consists in a resistive bolometer pixel associated with its conditioning and feedback electronics. Within the smart IR retina configuration, the processing means are shared between the 3D bolometers of the same line, and lines are stacked to form the imaging system.

Such 3D design presents an additional advantage compared to current devices that is an improved spatial resolution. Thermal infrared detectors since they are fabricated over the same substrate in the classical fabrication process are not thermally independent. As a consequence, the thermal diffusion between the pixels reduces the spatial resolution of the imaging system. In the case of the proposed structure the sensing area of the 3D pixel are physically isolated by empty spaces and each pixel will be thermally regulated, therefore the thermal diffusion between pixel will be avoided and then the spatial resolution enhanced compared to current devices.

Fig. 20. Smart IR retina. Several 3D bolometer pixels are associated with a common deported digital core on a line, here 4 or 6 3D pixels illustrated per line. Several lines, here 5, are stacked vertically to form the smart IR retina.

The fabrication technique is fully compatible with standard integrated circuit manufacturing methods enabling all the signal processing (conditioning, control and feedback) circuitry to be combined on the same chip with the sensor.

Even if the geometry principle and the fabrication technology are established, the fabrication of such monolithic device is still challenging and will require several years of development, but the work investment is worth.

5. Conclusion

The capacitively coupled electrical substitution technique is used to implement smart functions for uncooled resistive bolometers. Top-simulations and macro-scale experiments enable to derive the expected performance and functionalities of monolithic smart bolometer. This work takes place at a time where the microfabrication technologies and smart function integration converge to give birth to what could be a technological breakthrough in the field of infrared imaging.

"A rose with a microcontroller would be a smart rose".
-Randy Frank

6. Acknowledgment

This work was and is supported in part by the JSPS-Bridge program of the Japan Society for the Promotion of Science under No. BR-100201 and by the P-SOC program of the INS2I department of French CNRS under No. BFC-78518.

7. References

Allègre, G; Guillet, B; Robbes, D; Méchin, L; Lebargy, S; Nicoletti, S. (2007). A room temperature Si3N4/SiO2 membrane-type electrical substitution radiometer using thin film platinum thermometers, *Measurement Science and Technology*, Vol.18, No.1, pp 183-189.

Chen, X; Yi, X; Yang, Y; Li, Y. (2006). A new CMOS readout circuit for VO2-based uncooled focal plane arrays, *International Journal of Infrared and Millimeter Waves*, Vol.27, pp 1281-1291

Chapuis, Y.-A.; Zhou, L.; Fujita, H. & Hervé, Y. (2008). Multi-domain simulation using VHDL-AMS for distributed MEMS in functional environment: Case of a 2D air-jet micromanipulator, *Sensors and Actuators A: Physical*, Vol.148, No.1, pp 224-238

Denoual, M.; Delaunay, S. & Robbes, D. (2009). Bolometer with heat feedback, International patent, WO/2009/034066

Denoual, M.; Allègre, G.; Delaunay, S. & Robbes, D. (2009). Capacitively coupled electrical substitution for resistive bolometer enhancement, *Measurement Science and Technology*, Vol.20, DOI: 015105

Denoual, M.; Lebargy, S. & Allègre, G. (2010). Digital implementation of the capacitively coupled electrical substitution for resistive bolometers, *Measurement Science and Technology*, Vol.21, DOI: 015205

Denoual, M. & Allègre, G. (2010). Operating Uncooled Resistive Bolometers in a Closed-Loop Mode In: *Bolometers: Theory, Types and Applications*, Torrence M. Walcott Editor, NOVA Science publisher, ISBN: 978-1-61728-289-8

Denoual, M.; Delaunay, S.; Lebargy, S. & Allègre, G. (2010). CCES: A new configuration for electrical substitution for bolometers In: *Metrology and Instrumentation Selected Topics*, Sebastian Yuri Cavalcanti Catunda Editor, EDUFCG Campina Grande, ISBN: 978-85-8011-028-2

Denoual, M. & Attia, P. (2011). Standard VHDL Modeling and Top-Simulation for the Development of an Integrated Smart-Bolometer, Proceedings of SensorDevices 2011 2nd International Conference on Sensor Device Technologies and Applications, Nice, France, August 21-27, 2011

Frank, R. (2000). Understanding smart sensors, Artech House Inc. publisher, ISBN: 0-89006-311-7

Freire, R. C. S.; Catunda, S. Y. C. & Luciano, B. A. (2009). Applications of Thermoresistive Sensors Using the Electric Equivalence Principle, *IEEE Transactions on Instrumentation and Measurement*, Vol.58, No.6, pp 1823-1830

Galeazzi, M. (1998). An external electronic feedback system applied to a cryogenic micro-calorimeter, *Review of Scientific Instruments*, Vol.69, No.5, pp 2017-2023

Hirose, K.; Shiraishi, F.; Mita, Y. (2007). A simultaneous vertical and horizontal self-patterning method for deep three-dimensional microstructures, *Journal of Micromechanics and Microengineering*, Vol.17, S68–76

Joens, H.; Aslam, S.; Lakew, B. (2003), Bolometer simulation using SPICE, *TDW 2003, International Workshop on Thermal Detectors for Space Based Planetary, Solar, and Earth Science Applications*, Washington, DC USA, June 19-20, 2003

Kruse, P.W. (2001). Uncooled Thermal Imaging: Arrays, Systems and Applications, *SPIE Optical Engineering Press*, Bellingham, WA, ISBN: 0-8194-4122-8

Landau, I.D. (1988). Identification et commande des systèmes, *Traité des Nouvelles Technologies Hermés*, ISBN : 2-86601-119-8

Lee, I.I. (2010). A new readout integrated circuit for long-wavelength IR FPA, *Infrared Physics and Technology*, Vol.53, pp 140-145

Ljung, L.L. (1999). System identification – Theory for the User, *Prentice Hall PTR*, ISBN: 0-13-656695-2

Lv, J; Jiang, Y.D.; Zhang, D.L.; Zhou, Y. (2010). An ultra low noise readout integrated circuit for uncooled microbolometers, *Analog Integrated Circuits and Signal Processing*, Vol.63, pp 489-194

Mita, Y.; Kubota, M.; Harada, T.; Marty, F.; Saadany, B.; Bourouina, T. and Shibata, T. (2006). Contour lithography methods for DRIE fabrication of nanometer-millimetre-scale coexisting Microsystems, *Journal of Micromechanics and Microengineering*, Vol.16, S135-S141

Rice, J.P.; Lorentz, S.R.; Datla, R.U.; Vale, L.R.; Rudman, D.A.; Sing, M.L.C., & Robbes, D. (1998). Active cavity absolute radiometer based on high-Tc superconductors, *Metrologia*, Vol. 35, pp. 289-293

Rice, J.P. (2000). An electrically substituted bolometer as a transfer-standard detector, *Metrologia*, Vol.37, pp 433-436

Richards, P.L. (1994). Bolometers for infrared and millimeter waves, *Journal of Applied Physics*, Vol.76, pp 1–24

Rousseau, E.; Siria, A.; Jourdan, G.; Volz, S.; Comin, F.; Chevrier, J. & Greffet, J-J. (2009). Radiative heat transfer at the nanoscale, *Nature Photonics Letters*, DOI: 10.1038

Shen, S.; Narayanaswamy, A. & Chen., G. (2009). Surface Phonon Polaritons Mediated Energy Transfer between Nanoscale Gaps, *Nano Letters*, Vol.9, No.8, pp 2909-2913

Schubert, M. (1999). Mixed-Sginal Event-Driven Simulation of a Phase-Locked Loop, *IEEE International Behavioral Modeling and Simulation Conference*, Orlando, Florida, USA, October 4-6, 1999

Staszewski, R.B.; Fernando, C.; Balsara, P.T. (2005). Event-Driven Simulation and modeling of phase noise of an RF oscillator, *IEEE Transactions on Circuits and Systems*, 2005, Vol.52, No.4, pp 723-733

Texas Instrument. (2011). Infrared Thermopile Sensor in Chip-Scale Package, July 27, focus.ti.com/docs/prod/folders/print/tmp006.html

Williams, C.D.H. (1990). An appraisal of the noise performance of constant temperature bolometric detector systems, *Measurement Science and Technolology*, Vol.1, pp 322-328

Yole Développement, (2010). Uncooled IR Cameras & Detectors for Thermography and Vision, *Technologies & Market Report*, June

Zhuang, J.; Du, Q.; Kwaniewski, T. (2006). Event-Driven Modeling and Simulation of a Digital PLL, *IEEE International Behavioral Modeling and Simulation Conference*, San José, California, September 14-15, 2006, pp. 67-72

Detection of Terahertz Radiation from Submicron Plasma Waves Transistors

Y.M. Meziani[1], E. Garcia[1], J. Calvo[1], E. Diez[1], E. Velazquez[1],
K. Fobelets[2] and W. Knap[3]

[1]*Facultad de Ciencias, Universidad de Salamanca, Salamanca*
[2]*Department of Electrical and Electronic Engineering, Imperial College London, London*
[3]*LC2 Laboratory, Université Montpellier 2 & CNRS, Montpellier*
[1]*Spain*
[2]*United Kingdom*
[3]*France*

1. Introduction

Terahertz (THz) rays are located in the spectral region $\sim$ 0.1-10 THz ($\sim$ 3 mm - 30 μm, 3 cm^{-1} - 300 cm^{-1}) between the microwave and the infrared portion of the electromagnetic spectrum. Detection and emission of terahertz radiation bears great potential in medical diagnostics, product quality control and security screening. The attractive features of THz radiation for applications are: (i) They are transmitted by cloths and most packaging materials such as paper or plastics (ii) Many substances have "fingerprint" spectra in the THz range (iii) Due to its low photon energy (about one million times less than X-rays), THz radiation is non-ionizing and therefore not dangerous for human beings. These properties make THz systems a promising tool for different types of applications wherever detection and identification of hidden threats is the task and when human beings are to be scanned. When T-rays strike an object some of the rays pass through and others bounce back. A detector can measure the time it takes for rays to bounce back from different layers of an object. From that information, computers can produce a three dimensional image of the objects internal structure. The detectors can also identify the color shift of the reflecting rays. Every chemical and material has a unique color signature (fingerprint), so THz systems can determine not just what an object looks like, but what it's made of. THz rays may have dozens of practical applications, from medical imaging to wireless communications (Tonouchi, 2007). New terahertz system for scanning mails is under trial in Japan (Hoshina et al., 2009). Kawase et al. has demonstrated non-destructive terahertz imaging of illicit drugs (Kawase et al., 2003). An excellent review highlighting the importance of terahertz range in different types of applications from astronomy to spectroscopy can be found in Siegel (Siegel, 2002) and references in.

Terahertz detectors that rely on quantum transitions require cryogenic temperatures, since

the thermal energy ($k_B T$) needs to be smaller than the quantum transition energy (1 THz$\sim$4 meV~50 K) to avoid thermal saturation. Development of new detectors that can operate at room temperature is of big interest for terahertz technology. Dipole photoconductive antennas were developed first and are nowadays widely used in terahertz spectroscopy system for both detection and emission of terahertz pulses (Cheville, 2008; Hoffmann & Fülöp, 2011; Smith et al., 1988). The photoconductive antenna consists of a semiconductor material (GaAs, Low temperature-grown GaAs or silicon on saphire SOS) with a short carrier lifetime and an electrode structure with a gap of 10 μm. The electrode are biased to a voltage of the order of 10-50 V, resulting in high electric field strength (few KV cm^{-1}) across the switch. When the laser pulse hits the biased gap, free carriers are generated, and subsequently accelerated by the electric field. The rapid change in polarization induced by the ultrafast accelerations of the carriers generates a sub-picosecond electromagnetic pulse that partially propagates along the electrodes and, partially, is emitted to free space. The same physical phenomena is used to detect terahertz pulses using photoconductively gated antennas. The antenna is gated on and off by the optical pulse. Only when the laser pulse generates carriers the current flows in the direction of the THz electric field. The electric field of the terahertz wave can be measured as a function of time by scanning the time delay between the narrow gate laser pulse and the THz pulse.

Oscillations of the plasma waves in the channel of sub-micron transistors is one of promising tools for detection of terahertz radiations at room temperature. They present many advantages: low cost, small size, room temperature operation, and tuning of the resonant frequency by the gate voltage. The interest in the applications of plasma wave devices in the THz range started at the beginning of 90's with a pioneering theoretical work of Dyakonov and Shur (Dyakonov & Shur, 1996) who predicted that nonlinear properties of the two-dimensional (2D) plasma in the sub-micron transistor channel can be used for detection of THz radiation. Experimental investigations has been then conducted on different types of transistors demonstrating their capabilities for detection of terahertz radiation. Resonant detection from GaAs/AlGaAs FETs[1] was first reported by Knap et al. (Knap, Deng, Rumyantsev, Lü, Shur, Saylor & Brunel, 2002) at 8 K. Later, they reported on room temperature non-resonant detection (Knap, Kachorovskii, Deng, Rumyantsev, Lü, Gaska, Shur, Simin, Hu, Asif Khan, Saylor & C. Brunel, 2002). In 2004, it was demonstrated for the first time room temperature non resonant detection from silicon field-effect transistors (Knap et al., 2004) where the responsivity was estimated at around 200 V/W and the Noise Equivalent Power (NEP) at around 1 pW/$\sqrt{Hz}$ (Meziani et al., 2006; Tauk et al., 2006). THz imaging based on CMOS technology has been reported by different groups (Öjefors et al., 2009; Schuster et al., 2011). Recently, a responsivity of 80 KV/W and a NEP of 300 pW/$\sqrt{Hz}$ were reported using an array of Si-MOSFET processed by 0.25 μm CMOS technology as well as imaging at 0.65 THz (Lisauskas et al., 2009).

Here, we report on the detection of terahertz radiation by strained Si/Si$_{0.6}$Ge$_{0.4}$ n-MODFETs transistors. In the second section, we introduce the plasma wave oscillation under the theory of Dyakonov and Shur and we discuss both resonant and nonresonant detection cases. The third section describes the strained silicon devices and in the fourth one we present and discuss the observed resonant and non resonant detection from our devices and we explain these detections as due to the oscillations of the plasma waves in the channel. The last section

---

[1] FET: Field Effect Transistor

shows the capabilities of those devices in real applications where terahertz imaging using the strained Silicon devices are presented. All those results demonstrate the ability of plasma wave devices to be used in applications whenever detection of THz (0.1-10 THz) radiation is needed.

2. Principle of detection of terahertz radiation

When the electron plasma in the channel of a field-effect transistor is excited by an external electromagnetic radiation the induced ac electric field can be converted into measurable dc voltage (signal) via a nonlinear conversion mechanism. This signal has a resonant dependence on the incoming radiation with maxima at the plasma oscillation frequency, ω_0 and its odd harmonics Dyakonov & Shur (1995; 2001) $\omega_N = (1 + 2N)\omega_0$, where:

$$\omega_0 = \frac{\pi s}{2L} \tag{1}$$

L is the gate length, and s the plasma waves velocity which depends on the carrier density in the channel n_s, and on the gate-to-channel capacitance per unit area C :

$$s = \sqrt{\frac{e^2 n_s}{mC}} \tag{2}$$

where e is the absolute value of the electron charge and m is the electron effective mass. The surface carrier concentration (n_s) in the channel is related to the gate-to-channel voltage swing or overdrive voltage (U_0) by (Dyakonov & Shur, 2001):

$$n_s = \frac{CU_0}{e} \tag{3}$$

Here $U_0 = U_g - U_{th}$, U_g is the gate-to-channel voltage, and U_{th} is the threshold voltage at which the channel is completely depleted. Note that Eq. 3 is valid as long as the scale of the spacial variation of $U(x)$ is larger than the gate-to-channel separation (the gradual channel approximation). From Eqs. 1, 2, and 3, the fundamental plasma frequency can then be roughly expressed by the relation:

$$f_0 = \frac{1}{4L} \sqrt{\frac{eU_0}{m}} \tag{4}$$

This relation lead to two important consequences: (i) a submicron FET can operate as a terahertz detector (ii) the resonant frequency can be tuned by the gate bias.

The equations describing the 2D plasmon are the relationship between the surface carrier concentration and the swing voltage (Eq. 3), the equation of motion (Eq. 5), and the continuity equation (Eq. 6). The equation of motion (the Euler equation) is given by (Dyakonov & Shur, 2001):

$$\frac{\partial v}{\partial t} + \frac{e}{m}\frac{\partial U_0}{\partial x} + \frac{v}{\tau} = 0 \tag{5}$$

where $\partial U_0/\partial x$ is the longitudinal electric field in the channel, $v(x,t)$ is the local electron velocity, the last term is the viscosity and accounts for electronic collisions with phonons and/or impurities, and τ is the relaxation time. Equation 5 has to be solved together with the continuity equation which can be written as:

$$\frac{\partial U_0}{\partial t} + \frac{\partial (U_0 v)}{\partial x} = 0 \tag{6}$$

According to Dyakonov and Shur (Dyakonov & Shur, 1996; 2001), the solution of those equations under the boundary conditions of common-source/open-drain is given by:

$$\Delta U = \frac{U_a^2}{4U_0} f(\omega) \tag{7}$$

where ΔU is the source-to-drain voltage induced by the incoming radiation which is approximated by $U_a \cos(\omega t)$ (Dyakonov & Shur, 1996) and:

$$f(\omega) = 1 + \beta - \frac{1 + \beta \cos(2k_0' L)}{\sinh^2(k_0'' L) + \cos^2(k_0' L)} \tag{8}$$

Here

$$\beta = \frac{2\omega\tau}{\sqrt{1 + (\omega\tau)^2}} \tag{9}$$

$$k_0' = \frac{\omega}{s}\sqrt{\frac{(1 + (\omega\tau)^{-2})^{1/2} + 1}{2}} \tag{10}$$

$$k_0'' = \frac{\omega}{s}\sqrt{\frac{(1 + (\omega\tau)^{-2})^{1/2} - 1}{2}} \tag{11}$$

Equations 7 and 8 describe the response of the device as a THz detector for any frequency and gate length. The function $f(\omega)$ depends on two dimensionless parameters: $\omega\tau$ and $s\tau/L$. Figure 1 shows $f(\omega)$ as a function of $\omega\tau$ for different values of $s\tau/L$. When $\omega\tau \gg 1$ and for submicron devices, such that $s\tau/L \gg 1$, $f(\omega)$ exhibit sharp resonances at the fundamental frequency and its odd harmonics (Fig. 1(a) and (b)). In this case the damping of the plasma waves excited by incoming radiation is small and the device exhibits a resonance detection mode. However, when $\omega\tau \ll 1$, the plasma oscillations are overdamped. For a long device, the oscillations excited at the source by the incoming radiation do not reach the drain because of the damping. The boundary conditions at the drain are irrelevant in this case, and the response does not depend on L. As it can be seen in Fig. 1(c) and (d), $f(\omega)$ changes from $f = 1$ for $\omega\tau \ll 1$ to $f = 3$ for $\omega\tau \gg 1$, where we also see how at very small values of $\omega\tau$ the condition of a long sample is violated and f tends to zero. In both cases, a long channel acts as a broadband detector of electromagnetic radiation. Underdamped $\omega\tau \gg 1$ or overdamped $\omega\tau \ll 1$ plasma waves decay near the source end of the channel, leading to a dc voltage induced between drain and source.

3. Devices description

The epistructure of the MODFET was grown by molecular beam epitaxy (MBE) on a thick relaxed SiGe virtual substrate grown by low-energy plasma-enhanced chemical vapor deposition (LEPECVD) over a p-doped conventional Si wafer. The final Ge molar

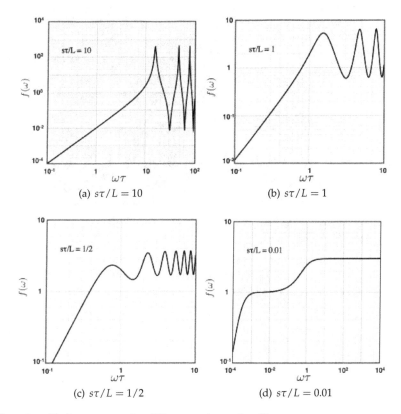

Fig. 1. Function $f(\omega)$ versus $\omega\tau$ for different values of $s\tau/L$.

concentration in the virtual substrate was $x_{Ge} = 0.45$. The device had a 8 nm tensile strained (in terms of biaxial deformation) Si channel sandwiched between two heavily doped SiGe electron supply layers to generate a high carrier density in the strained-Si quantum well (Rumyantsev et al., 2008). The ohmic contacts were not self-aligned. Two transistors were measured with different gate lengths (150 nm and 250 nm). The gate width and the source-to-drain length are 30 μm and 1μm respectively. Figure 2 shows an image of three devices with (50 nm, 150 nm, and 250 nm) gate length with common drain and different source pads.

4. Results and discussion

4.1 Resonant detection

Figure 3(a) shows the photoresponse signal as a function of the gate bias for strained Silicon transistor excited with electromagnetic radiation at 323 and 360 GHz. The gate length is 250 nm and the measurement was performed at 4 K. A clear shift of the response's peak to higher gate voltage values with the frequency of the excitation was observed. The square and triangle symbols in Fig. 3(b) mark the maximum of signals at 363 and 323 GHz respectively. According

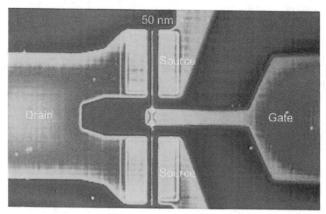

Fig. 2. Image of strained Silicon MODFET with 50 nm gate length. The image was captured using a 3D Laser Microscope Olympus LEXT OLS4000.

to Eq. 4, the resonant frequency as a function of the voltage swing is plotted for Lg=230nm Fig. 3(b) and the observed displacement towards higher values of the gate voltage is in good agreement with Dyakonov-Shur prediction.

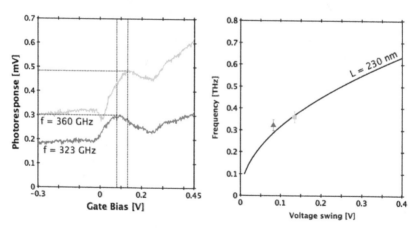

Fig. 3. (a) Photoresponse vs gate bias at two frequencies of the incoming radiation (323 and 360 GHz). (b) Resonance frequency vs swing voltage for Lg=230 nm gate lengths (200, 230, and 260 nm).

In the presence of a magnetic field, Eq. 7 can be written in the following form (Lifshits & Dyakonov, 2009):

$$\Delta U = \frac{1}{4}\frac{U_a^2}{U_0}\left[f(\beta) - \frac{d\gamma}{dn}\frac{n}{\gamma}g(\beta)\right] \qquad (12)$$

where $\beta = \frac{\omega_c}{\omega}$ (ω_c and ω are the cyclotron resonance and the incoming radiation frequency respectively) is a normalized cyclotron resonance. Dependence of the photoresponse on the magnetic field and the radiation frequency is described by the functions $f(\beta)$ and $g(\beta)$:

$$f(\beta) = 1 + \frac{1+F}{\sqrt{(\alpha^2 + F^2)}} \tag{13}$$

$$g(\beta) = \frac{1+F}{2}\left(1 + \frac{1+F}{\sqrt{(\alpha^2 + F^2)}}\right) \tag{14}$$

$$F = \frac{1 + \alpha^2 - \beta^2}{1 + \alpha^2 + \beta^2} \quad and \quad \alpha = (\omega\tau)^{-1} \tag{15}$$

The γ factor is an oscillating function of the magnetic field and the electron concentration that can be correlated with the Shubnikov de Haas oscillations. The second term in the right-hand side of Eq. 12, proportional to $d\gamma/dn$, is an oscillating function of the gate voltage and the magnetic field. The derivative in Eq. 12 induces a shift of $\pi/2$ between the SdH oscillations and the photoresponse. This result has been previously observed on InGaAs FETs (Klimenko et al., 2010) and here we report the same phenomenon on Si/SiGe MODFET (Fig. 4) for the first time.

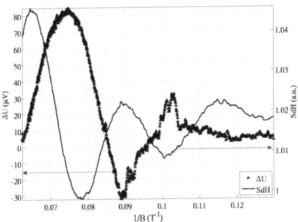

Fig. 4. Photoresponse (dashed line with triangles) and the oscillating part of the magnetoresistance (solid line) vs the inverse of the magnetic field.

4.2 Non resonant detection

The devices were excited at room temperature by a terahertz wave parametric oscillator (TPO) pulsed laser at 1.5 THz (Minamide et al., 2009). The TPO system consists of three mirrors and a MgO:LiNbO$_3$ crystal under non-collinear phase-matching conditions. It can emit monochromatic THz-waves over a wide tunable frequency range from 0.4 THz to 2.8 THz with a narrow line-width lower than 100 MHz. The output power of the laser was 6 nJ/pulse for the range 1.3-1.6 THz and the repetition rate was 500 Hz. The radiation was coupled to the dice via the metallization pads. The source terminal was grounded. The radiation intensity was modulated by a mechanical chopper at 1.29 KHz and the induced photoresponse signal is measured by using the lock-in amplifier technique (Fig. 5). A wire grid polarizer was used to polarize the light in parallel with the channel. The principle of operation is as follows: when the device is excited by an external electromagnetic radiation the induced ac electric fields can be converted into a measurable dc signal via a nonlinear conversion mechanism. This will be referred hereafter as the photoresponse signal.

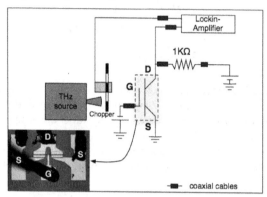

Fig. 5. Schematic description of the experimental setup of the detection of THz radiation.

Figure 6 shows the photoresponse signal as a function of the gate bias for two devices with different gate lengths (L_G=150 and 250 nm) excited by 1.5 THz radiation at room temperature (Meziani et al., 2011). The signal is presented for different drain currents: 20, 50 and 100 μA. The intensity increases with the drain current as predicted by the theory (Veksler et al., 2006) and a maximum is observed around the threshold voltage. The same behavior has been recently reported by Elkhatib et al. (2011) demonstrating that the response signal linearly increases with the drain current (or drain-to-source voltage) and the responsivity might reach high values within the saturation regime. The observed signal intensity is low because no parabolic mirror was used to focus the beam and also, as reported in Rumyantsev et al. (2008), the response intensity at terahertz frequencies (over 1 THz) is considerably smaller than at sub-terahertz ones (below 1 THz). The responsivity was estimated to be 20 V/J/Pulse at I_{DS} = 100 μA. A non-resonant response has been reported for Si-FET Knap et al. (2004) and it was related to a low value of the quality factor i.e. low carrier mobility in the device. The quality factor was found to be around 1.2 for μ=1355 cm^2/V.s and for f=1.5 THz. Rumyantsev et al. (2008) obtained on similar devices a maximum value of the photoresponse signal when the beam focus was away from the transistor. This is a proof of low coupling of the THz radiation to the device. To increase the efficiency toward high non-resonant signal and possible resonant detection, new designs are under consideration: array of transistors, devices with larger pads, and grating devices Popov et al. (2011a).

5. Terahertz imaging

First, a strained Silicon MODFET transistor with a gate length of 50 nm was subjected at room temperature to electromagnetic radiation at 292 GHz from a Gunn diode (Fig. 5). Figure 7 shows the photoresponse signal as a function of the gate bias. A maximum signal is observed around the threshold voltage ($\sim$ $-0.84V$). This behavior has been reported earlier (Knap et al., 2004; Rumyantsev et al., 2008) and explained as non-resonant detection. However, the

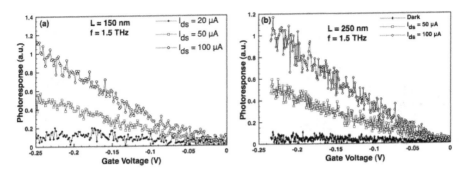

Fig. 6. Photoresponse vs gate bias at 1.5 THz and for two devices lengths: (a) 150 nm and (b) 250 nm.

signal to noise ratio shown here is high enough to allow the use of this devices for different applications like terahertz imaging. The noise equivalent power of this device is estimated around 1 pW/$\sqrt{Hz}$ in the same order of CMOS devices. Subsequently, the device was used

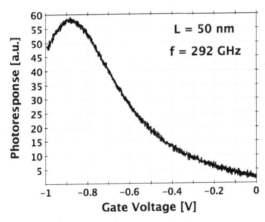

Fig. 7. Measured photoresponse of the device as function of the gate voltage. The curve shows a maximum near to the threshold voltage of the MODFET.

as a detector in a terahertz imaging system. Figure 8 describes schematically the terahertz imaging system. A Gunn diode was used as a source of terahertz radiation with f = 292 GHz. The radiation is collimated and focused by off-axis parabolic mirrors and a visible red LED in combination with an indium tin oxide (ITO) mirror are used for the alignment of the THz beam. The incident THz light is mechanically chopped at 333 Hz and the photo-induced drain-to-source voltage ΔU is measured using a lock-in technique. All measurements were done at room temperature. More information about the THz imaging system can be found in Schuster et al. (2011). The gate bias (V_g) was fixed at a value close to the threshold voltage to obtain optimum signal for imaging and high values of the signal to noise ratio. Figure 9 shows both visible (left) and terahertz image (right) of a box with a hidden mirror inside. The resolution of the image is 17500 pixels within x axis and of 120000 pixels within y axis.

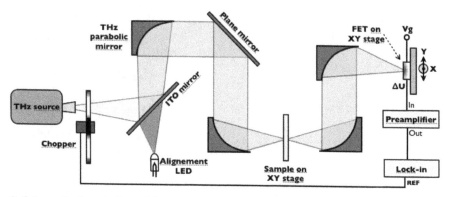

Fig. 8. Schematic description of the experimental setup of the imaging system.

Terahertz image of a VISA card is also shown in Fig. 10. The resolution of the terahertz image is 23000x110000 pixels in XxY axis. Those images demonstrate the capability of such devices in compact terahertz imaging systems and that they could play an important role in other THz applications. Those systems are inexpensive, compact, operate at room temperature and can be monolithically integrated along with Si-circuitry.

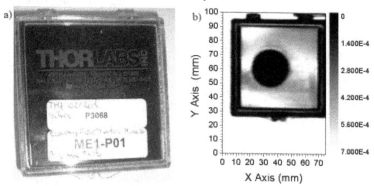

Fig. 9. (a) Visible and (b) terahertz image of a plastic box with a hidden mirror inside.

6. Future directions

Increasing the responsivity and the noise equivalent power are the main goals to improve the performance of the next generation of plasma wave THz detectors. This may be achieved by using new materials (like Graphene) and also by using new designs for a better coupling of the terahertz radiation into the transistor. Recently, different groups have reported record values of the NEP from different well designed devices. High responsivity (200 V/W) and low noise equivalent power (1 pW/$\sqrt{Hz}$) were observed from a Si-MOSFETs with different gate lengths (120 and 300 nm), Tauk et al. (2006). Terahertz imaging based on CMOS technology has been recently reported by different groups: Lisauskas et al. (2009); Schuster et al. (2011; 2010). A 3x5 Si MOSFET focal-plane array processed using conventional 0.25 μm CMOS technology was used by Lisauskas et al. (2009) for imaging at 0.65 THz. Responsivity of 80 KV/W and a NEP

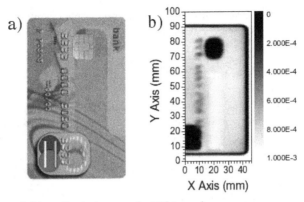

Fig. 10. (a) Visible and (b) terahertz image of a VISA card.

of 300 pW/$\sqrt{Hz}$ was obtained. Recently, terahertz imaging was performed using a nMOS field effect transistor with an integrated bow-tie coupling antenna with a responsivity above 5 kV/W and a noise equivalent power below 10 pW/$\sqrt{Hz}$, Schuster et al. (2011). Asymmetric doubly interdigitated grating gates structures with theoretical values of the responsivity in excess of 5 KV/W are under consideration (Otsuji et al., 3 December 2010; Popov et al., 2011b). Recently, a responsitivity up to 2.2 KV/W at 1 THz has been reported by using asymetric grating gate structures Otsuji (November 2011).

7. Conclusion

We demonstrated the capability of submicron Strained Si/Si$_{0.6}$Ge$_{0.4}$ n-MODFETs transistors for the detection of terahertz radiation. Resonant detection has been observed at low temperature and related to the plasma oscillations in the channel. A broadband (non resonant) detection has also been reported using the devices under continuous and pulsed excitation. We have shown the capabilities of those transistors in real terahertz imaging at room temperature. We believe that strained-Si transistors could play an important role in different terahertz applications in the near future.

8. Acknowledgment

Authors from Salamanca University acknowledge the financial help from the Ministry of Science and Innovation (MICINN) through the projects PPT-120000-2009-4, PCT-420000-2010-008 and TEC2008-02281 and Junta de Castilla y León (grants Numbers SA061A09 and SA049A10). Y.M.M. thanks the Ramon y Cajal Program for support. The authors from Montpellier University acknowledge the CNRS GDR-I project "Semiconductors Sources and Detectors of THz radiation" and the Region of Languedoc-Roussillon through the "Terahertz Platform" project.

9. References

Cheville, R. A. (2008). Terahertz time-domain spectroscopy with photoconductive antennas, *in* S. L. Dexheimer (ed.), *Terahertz Spectroscopy*, Taylor and Francis Group, UK, pp. 1–39.

Dyakonov, M. & Shur, M. (1996). Detection, mixing, and frequency multiplication of terahertz radiation by two-dimensional electronic fluid, *IEEE Transactions on Electron Devices* 43(3): 380 –387.

Dyakonov, M. & Shur, M. S. (1995). *Proceedings of the 22nd International Symposium on GaAs and Related Compounds*, Institute Conference Series No 145, Cheju, Korea, Chap. 5, pp. 785–790 (1996).

Dyakonov, M. & Shur, M. S. (2001). Plasma wave electronics for terahertz applications, *in* R. Miles, P. Harrison & D. Lippens (eds), *Terahertz Sources and Systems*, Kluwer Academic Publishers, Netherlands, pp. 187–207.

Elkhatib, T. A., Kachorovskii, V. Y., Stillman, W. J., Rumyantsev, S., Zhang, X.-C. & Shur, M. S. (2011). Terahertz response of field-effect transistors in saturation regime, *Appl. Phys. Lett.* 98(24): 243505.
URL: *http://dx.doi.org/doi/10.1063/1.3584137*

Hoffmann, M. & Fülöp, J. (2011). Intense ultrashort terahertz pulses: generation and applications, *Journal of Physics D: Applied Physics* 44: 083001.

Hoshina, H., Sasaki, Y., Hayashi, A., Otani, C. & Kawase, K. (2009). Noninvasive mail inspection using terahertz radiation, *SPIE Newsroom* .

Kawase, K., Ogawa, Y., Watanabe, Y. & Inoue, H. (2003). Non-destructive terahertz imaging of illicit drugs using spectral fingerprints, *Opt. Express* 11(20): 2549–2554.
URL: *http://www.opticsexpress.org/abstract.cfm?URI=oe-11-20-2549*

Klimenko, O. A., Mityagin, Y. A., Videlier, H., Teppe, F., Dyakonova, N. V., Consejo, C., Bollaert, S., Murzin, V. N. & Knap, W. (2010). Terahertz response of InGaAs field effect transistors in quantizing magnetic fields, *Appl. Phys. Lett.* 97(2): 022111.
URL: *http://dx.doi.org/doi/10.1063/1.3462072*

Knap, W., Deng, Y., Rumyantsev, S., Lü, J.-Q., Shur, M. S., Saylor, C. A. & Brunel, L. C. (2002). Resonant detection of subterahertz radiation by plasma waves in a submicron field-effect transistor, *Appl. Phys. Lett.* 80(18): 3433–3435.
URL: *http://dx.doi.org/doi/10.1063/1.1473685*

Knap, W., Kachorovskii, V., Deng, Y., Rumyantsev, S., Lü, J.-Q., Gaska, R., Shur, M. S., Simin, G., Hu, X., Asif Khan, M., Saylor, C. A. & C. Brunel, L. (2002). Nonresonant detection of terahertz radiation in field effect transistors, *J. Appl. Phys.* 91(11): 9346–9353.
URL: *http://dx.doi.org/doi/10.1063/1.1468257*

Knap, W., Teppe, F., Meziani, Y., Dyakonova, N., Lusakowski, J., Boeuf, F., Skotnicki, T., Maude, D., Rumyantsev, S. & Shur, M. S. (2004). Plasma wave detection of sub-terahertz and terahertz radiation by silicon field-effect transistors, *Appl. Phys. Lett.* 85(4): 675–677.
URL: *http://dx.doi.org/doi/10.1063/1.1775034*

Lifshits, M. B. & Dyakonov, M. I. (2009). Photovoltaic effect in a gated two-dimensional electron gas in magnetic field, *Phys. Rev. B* 80: 121304.
URL: *http://link.aps.org/doi/10.1103/PhysRevB.80.121304*

Lisauskas, A., Pfeiffer, U., Ëjefors, E., Bolşvar, P. H., Glaab, D. & Roskos, H. G. (2009). Rational design of high-responsivity detectors of terahertz radiation based on distributed self-mixing in silicon field-effect transistors, *J. Appl. Phys.* 105(11): 114511.
URL: *http://dx.doi.org/doi/10.1063/1.3140611*

Meziani, Y. M., Garcia, E., Velazquez, E., Diez, E., Moutaouakil, A. E., Otsuji, T. & Fobelets, K. (2011). Strained silicon modulation field-effect transistor as a new sensor of terahertz

radiation, *Semiconductor Science and Technology* 26(10): 105006.
URL: *http://stacks.iop.org/0268-1242/26/i=10/a=105006*

Meziani, Y. M., Lusakowski, J., Dyakonova, N., Knap, W., Seliuta, D., Sirmulis, E., Devenson, J., Valusis, G., Boeuf, F. & Skotnicki, T. (2006). Non resonant response to terahertz radiation by submicron CMOS transistors, *IEICE TRANSACTIONS on Electronics* E89-C: 993–998.

Minamide, H., Ikari, T. & Ito, H. (2009). Frequency-agile terahertz-wave parametric oscillator in a ring-cavity configuration, *Rev. Sci. Instrum.* 80(12): 123104.
URL: *http://dx.doi.org/doi/10.1063/1.3271039*

Öjefors, E., Lisauskas, A., Glaab, D., Roskos, H. & Pfeiffer, U. (2009). Terahertz imaging detectors in CMOS technology, *Journal of Infrared, Millimeter and Terahertz Waves* 30: 1269–1280. 10.1007/s10762-009-9569-4.
URL: *http://dx.doi.org/10.1007/s10762-009-9569-4*

Otsuji, T., Popov, V., Knap, W., Meziani, Y., Dyakonova, N., D. Coquillat, F. T., Fateev, D. & Perez, J. E. V. (3 December 2010). Terahertz apparatus, *Japanese Patent PCT/JP2010/007074*.

Otsuji, T. (2011). Emission and detection of terahertz radiation using two-dimensional electrons in III-V semiconductors and Graphene, Proceedings of the joint conference for International Symposium on Terahertz Nanoscience (Teranano2011) and Wokshop of International Terahertz Research Network (GDR-I THz 2011), Osaka University Nakanoshima Center, Osaka, Japan, pp. 133-136.

Popov, V., Fateev, D., Otsuji, T., Meziani, Y., Coquillat, D. & Knap, W. (2011a). Plasmonic detection of terahertz radiation in a double-grating-gate transistor structure with an asymmetric unit cell, *Proceedings of XV International Symposium "Nanophysics and Nanoelectronics"*, Vol. 1, Institute Conference Series No 145, Nizhnii Novgorod, Russia, pp. 121–122.

Popov, V., Fateev, D., Otsuji, T., Meziani, Y., Coquillat, D. & Knap, W. (2011b). Plasmonic terahertz detection by a double-grating-gate field-effect transistor structure with an asymmetric unit cell, *Submitted to Appl. Phys. Lett.*
URL: *http://arxiv.org/abs/1111.1807*

Rumyantsev, S. L., Fobelets, K., Veksler, D., Hackbarth, T. & Shur, M. S. (2008). Strained-si modulation doped field effect transistors as detectors of terahertz and sub-terahertz radiation, *Semiconductor Science and Technology* 23(10): 105001.
URL: *http://stacks.iop.org/0268-1242/23/i=10/a=105001*

Schuster, F., Coquillat, D., Videlier, H., Sakowicz, M., Teppe, F., Dussopt, L., Giffard, B., Skotnicki, T. & Knap, W. (2011). Broadband terahertz imaging with highly sensitive silicon CMOS detectors, *Opt. Express* 19(8): 7827–7832.
URL: *http://www.opticsexpress.org/abstract.cfm?URI=oe-19-8-7827*

Schuster, F., Sakowicz, M., Siligaris, A., Dussopt, L., Videlier, H., Coquillat, D., Teppe, F., Giffard, B., Dobroiu, A., Skotnicki, T. & Knap, W. (2010). THz imaging with low-cost 130 nm CMOS transistors, *Proc. SPIE* 7837(1): 783704.
URL: *http://dx.doi.org/doi/10.1117/12.864877*

Siegel, P. (2002). Terahertz technology, *IEEE Transactions on Microwave Theory and Techniques*, 50(3): 910–928.

Smith, P., Auston, D. & Nuss, M. (1988). Subpicosecond photoconducting dipole antennas, *IEEE Journal of Quantum Electronics* 24(2): 255–260.

Tauk, R., Teppe, F., Boubanga, S., Coquillat, D., Knap, W., Meziani, Y. M., Gallon, C., Boeuf, F., Skotnicki, T., Fenouillet-Beranger, C., Maude, D. K., Rumyantsev, S. & Shur, M. S. (2006). Plasma wave detection of terahertz radiation by silicon field effects transistors: Responsivity and noise equivalent power, *Appl. Phys. Lett.* 89(25): 253511.
URL: *http://dx.doi.org/doi/10.1063/1.2410215*

Tonouchi, M. (2007). Cutting-edge terahertz technology, *Nature Photonics* 1: 97–105.

Veksler, D., Teppe, F., Dmitriev, A. P., Kachorovskii, V. Y., Knap, W. & Shur, M. S. (2006). Detection of terahertz radiation in gated two-dimensional structures governed by dc current, *Phys. Rev. B* 73: 125328.
URL: *http://link.aps.org/doi/10.1103/PhysRevB.73.125328*

Permissions

The contributors of this book come from diverse backgrounds, making this book a truly international effort. This book will bring forth new frontiers with its revolutionizing research information and detailed analysis of the nascent developments around the world.

We would like to thank Prof. A. G. Unil Perera, for lending his expertise to make the book truly unique. He has played a crucial role in the development of this book. Without his invaluable contribution this book wouldn't have been possible. He has made vital efforts to compile up to date information on the varied aspects of this subject to make this book a valuable addition to the collection of many professionals and students.

This book was conceptualized with the vision of imparting up-to-date information and advanced data in this field. To ensure the same, a matchless editorial board was set up. Every individual on the board went through rigorous rounds of assessment to prove their worth. After which they invested a large part of their time researching and compiling the most relevant data for our readers. Conferences and sessions were held from time to time between the editorial board and the contributing authors to present the data in the most comprehensible form. The editorial team has worked tirelessly to provide valuable and valid information to help people across the globe.

Every chapter published in this book has been scrutinized by our experts. Their significance has been extensively debated. The topics covered herein carry significant findings which will fuel the growth of the discipline. They may even be implemented as practical applications or may be referred to as a beginning point for another development. Chapters in this book were first published by InTech; hereby published with permission under the Creative Commons Attribution License or equivalent.

The editorial board has been involved in producing this book since its inception. They have spent rigorous hours researching and exploring the diverse topics which have resulted in the successful publishing of this book. They have passed on their knowledge of decades through this book. To expedite this challenging task, the publisher supported the team at every step. A small team of assistant editors was also appointed to further simplify the editing procedure and attain best results for the readers.

Our editorial team has been hand-picked from every corner of the world. Their multi-ethnicity adds dynamic inputs to the discussions which result in innovative outcomes. These outcomes are then further discussed with the researchers and contributors who give their valuable feedback and opinion regarding the same. The feedback is then collaborated with the researches and they are edited in a comprehensive manner to aid the understanding of the subject.

Apart from the editorial board, the designing team has also invested a significant amount of their time in understanding the subject and creating the most relevant covers. They scrutinized every image to scout for the most suitable representation of the subject and create an appropriate cover for the book.

The publishing team has been involved in this book since its early stages. They were actively engaged in every process, be it collecting the data, connecting with the contributors or procuring relevant information. The team has been an ardent support to the editorial, designing and production team. Their endless efforts to recruit the best for this project, has resulted in the accomplishment of this book. They are a veteran in the field of academics and their pool of knowledge is as vast as their experience in printing. Their expertise and guidance has proved useful at every step. Their uncompromising quality standards have made this book an exceptional effort. Their encouragement from time to time has been an inspiration for everyone.

The publisher and the editorial board hope that this book will prove to be a valuable piece of knowledge for researchers, students, practitioners and scholars across the globe.

List of Contributors

Mario Moreno, Alfonso Torres and Andrey Kosarev
National Institute of Astrophysics, Optics and Electronics, INAOE, Mexico

Roberto Ambrosio
Universidad Autonoma de Ciudad Juarez, UACJ, Mexico

Henry H. Radamson
School of Information and Communication Technology, KTH Royal Institute of Technology, Kista, Sweden

M. Kolahdouz
Thin Film Laboratory, Electrical and Computer Engineering Department, University of Tehran, Tehran, Iran

Leonid S. Kuzmin
Chalmers University of Technology, Department of Micro technology and Nano science, Sweden

Lei Liu
Department of Electrical Engineering, University of Notre Dame, Notre Dame, IN, USA

Béla Szentpáli
Hungarian Academy of Sciences, Research Institute for Technical Physics and Materials Science, Hungary

Kumudni Tahiliani and Ratneshwar Jha
Institute for Plasma Research, India

Matthieu Denoual and Gilles Allègre
University of Caen Basse-Normandie, GREYC-ENSICAEN, France

Olivier de Sagazan
University of Rennes, IETR, France

Patrick Attia
NXP-semicoductors Caen, France

Y.M. Meziani, E. Garcia, J. Calvo, E. Diez and E. Velazquez
Facultad de Ciencias, Universidad de Salamanca, Salamanca, Spain

K. Fobelets
Department of Electrical and Electronic Engineering, Imperial College London, London, United Kingdom

W. Knap
LC2 Laboratory, Université Montpellier 2 & CNRS, Montpellier, France

9 781632 384591